Handbooks for the
Identification of British Insects

Vol. 10, Part 4a(i)

Editors: W. R. Dolling & R. R. Askew

TACHINID FLIES

DIPTERA: TACHINIDAE

By

Robert Belshaw

Department of Entomology
Natural History Museum
London SW7 5BD

1993
ROYAL ENTOMOLOGICAL SOCIETY OF LONDON

The aim of the Handbooks is to provide illustrated identification keys to the insects of Britain, together with concise morphological, biological and distributional information. The series also includes a *Check list of British Insects*.

Each Handbook should serve both as an introduction to a particular group of insects and as an identification manual.

Details of Handbooks currently available, and an order form, can be obtained from the Royal Entomological Society, 41 Queen's Gate, London SW7 5HR.

World List abbreviation: *Handbk Ident. Br. Insects*.

© Royal Entomological Society of London, 1993.

First published 1993 by the Royal Entomological Society of London, 41 Queen's Gate, London SW7 5HR.

Cover illustration © Steven Falk.

ISBN 0 901546 81 X

Printed by Dramrite Printers Limited, Southwark SE1 4PL

Contents

Abstract

An identification key is provided for the adults of British Tachinidae (Diptera). A summary of the biology and distribution of each species is given, including a list of British host records. The natural history of the family is reviewed briefly, with emphasis placed on the reproductive strategies and interactions with the host. A check list includes all old generic and specific names used in the British literature since 1928.

Acknowledgements

During preparation of this work, specimens were borrowed from the following institutions.

University Museum, Manchester.
Leicester Museum.
Hope Department of Entomology, Oxford.
Liverpool Museum.
National Museum of Wales, Cardiff.
Staatliches Museum für Naturkunde, Stuttgart.
Central Museum and Art Gallery, Bolton.
Zoologisk Museum, København.
Institut für Pflanzenschutzforschung, Eberswalde-Finow.

In addition, facilities and assistance were provided to examine the collections of Oxford, Manchester, Cardiff and Stuttgart.

Steven Falk provided a copy of his manuscript (courtesy of the Nature Conservancy Council). Tom Ford lent specimens from his private collection and provided a copy of a manuscript of his and Mark Shaw's. Stig Andersen and Peter Tschorsnig helped with many identification problems. Peter Tschorsnig and Dr B. Herting provided a draft of their (unpublished) Palaearctic Tachinidae host list and key to European species. Peter Chandler supplied additional data for the Species biology section. Mike Bloxham, Peter Hodge and Eric Philp tested drafts of the key. Paul Eggleton helped with the introduction. I am very grateful to all the above, and to the dipterists at the NHM, in particular to Brian Pitkin, Bruce Townsend and Nigel Wyatt. Above all I would like to acknowledge the contribution of Nigel Wyatt, whose curation of the NHM British Tachinidae collection forms the basis of this work.

This publication was researched and written at the Natural History Museum, London while in receipt of a 3 year Research Fellowship from the Royal Entomological Society of London. Additional financial support was received from the Systematics Association's Small Grants Fund.

Family biology

Introduction and life history

The Tachinidae form the second largest family of Diptera, with approximately 8,200 described species world-wide (Cantrell & Crosskey, 1989). They are well represented in all zoogeographical zones with 241 species listed here as British (although a few of these may be extinct in Britain).

All tachinids are parasitoids, their larvae developing within other arthropods, predominantly other insects. The majority of tachinid species attack Lepidoptera larvae but adults and immatures of many other orders are also attacked.

They are an important group economically, having been used extensively as biological control agents (see Grenier, 1988 for a review). In this respect they are comparable with the more important families of parasitoid Hymenoptera (Greathead, 1986).

The adults of most species are between 5 and 10mm in length and are found in most habitats: resting on foliage, feeding on nectar or honey-dew, or flying slowly in search of hosts. Mating often takes place at specific sites where the males assemble, such as the tops of isolated shrubs, hilltops, or the tips of prominent branches (Wood, 1987). Males of many species are found resting in sunlight on foliage or tree trunks, waiting to chase passing females. Male swarming behaviour has also been observed in several species, including *Carcelia puberula* (Herting, 1960) and *Siphona* spp. (Andersen, 1982).

Tachinids employ a variety of strategies in attacking their hosts: adult females may lay eggs directly on or in the host; active first-instar larvae may seek out the host and then bore their way in; or the female may lay eggs on the host food-plant which are then accidentally ingested by the host while it is feeding.

There are three larval instars. The mature tachinid larva usually leaves the host to pupate, either (in the case of exposed leaf-feeding hosts) in the soil or (in the case of concealed hosts) within the host gallery or leaf-roll. Some species attacking Lepidoptera larvae allow the host to form its protective cocoon before killing it and pupate within that. Species attacking adult Coleoptera often pupate within the host abdomen. In some species only a single individual completes its development in each host (= solitary), but in others more than one may do so (= gregarious). Often the number is determined by the size of the host.

The majority of species overwinter as puparia but some overwinter as larvae within the living host. The number of generations per year ranges from one to several depending on the species. Development times vary considerably between species and with temperature. Excluding overwintering, the larval and pupal stages both last for 1 to 3 weeks, and the adult stage for 1 to 2 months.

If a host individual is parasitised by more than one larva of a solitary species, these often fight using their mouthparts until only one remains. In contrast larvae of gregarious species do not appear to fight, but if too many are present the resulting adults may be reduced in size. If two gregarious species parasitise the same host individual the successful species is usually the one which entered the host first and/or develops faster (e.g. Godwin & Odell, 1984).

Several cases of different species of tachinid emerging from the same host individual have been recorded, e.g. *Compsilura concinnata* can successfully develop alongside other tachinid (pers. obs.) and hymenopteran parasitoids (Weseloh, 1983). This appears to result from the absence of fighting among larvae of gregarious species, and the failure of many species to consume their host completely.

Occasionally the host survives the emergence of the tachinid larva, especially if the host is much larger than the tachinid. There have been isolated cases of individual Lepidoptera larvae completing metamorphosis (Richards & Waloff, 1948; DeVries, 1984). In one study, 4 arctiid larvae, from which tachinids had emerged, completed their development and laid a normal complement of eggs (English-Loeb *et al.*, 1990). Individuals of a predacious pentatomid species (Hemiptera) have been observed to feed on the tachinid larva which emerged from them (Aldrich *et al.*, 1984). Eight adult female ants (*Lasius niger*), collected after their mating flight, survived the emergence of tachinid larvae for between 1 and 2 months. They even cared for the tachinid puparia as if they were their own brood; cleaning them and returning them to the nest if they were removed (Gösswald, 1949 cited in Herting, 1960).

Host groups

The range of arthropods attacked by British tachinids is summarised in table 1.

Table 1. Hosts of British Tachinidae (excluding some questionable records).

Host group	Attacked by
Lepidoptera larvae	
exposed on plants	103 species
concealed in plants	30 species
(leaf-rollers, stem-borers *etc.*)	
Sawfly larvae	14 species in total (including some polyphagous
(Hymenoptera: Symphyta)	species usually reared from Lepidoptera)
Tenthredinidae	*Myxexoristops*, some *Exorista, Hyalurgus lucidus, Bessa selecta, Meigenia mutabilis* and *Phyllomya volvulus.*
Cimbicidae	some *Phebellia*
Diprionidae	*Diplostichus janithrix*
Pamphilidae	*Myxexoristops blondeli*
Argidae	*Belida angelicae*
Coleoptera	32 species in total
adult Coccinellidae	*Medina*
adult Chrysomelidae	*Medina* and *Policheta unicolor*
larval Chrysomelidae	*Meigenia, Macquartia, Dufouria, Anthomyiopsis nigrisquamata* and *Cleonice callida*
adult Curculionidae	*Microsoma exigua* and possibly *Rondania fasciata*
larval Tenebrionidae	*Chetogena acuminata*
adult Carabidae	*Zaira cinerea* and *Freraea gagatea*
larval Carabidae	*Dinera grisescens*
larval Scarabaeidae	*Estheria, Dexia, Dexiosoma caninum, Dinera carinifrons* and *Prosena siberita*
larval Cerambycidae	*Billaea irrorata*
Tipulidae larvae	*Admontia, Siphona geniculata* and possibly
(Diptera)	*Erycilla ferruginea*
Earwigs	*Ocytata pallipes* and *Triarthria setipennis*
(Dermaptera)	
Centipedes	*Loewia* and *Eloceria delecta*
(Chilopoda: *Lithobius)*	
Heteropteran bugs	Phasiinae (20 species)
(Hemiptera)	

Exposed leaf-feeding Lepidoptera larvae are the group most commonly attacked, but concealed microlepidoptera larvae, sawfly larvae, Coleoptera (adults and larvae) and heteropteran bugs are also important host groups.

The majority of British species (where sufficient host records exist) show an association with a particular family or (more rarely) genus of host (see 'Species biology'). Others attack hosts in particular ecological situations, e.g. exposed Lepidoptera larvae on trees. Restriction to a single species of host (monophagy) is rare: among British species attacking Lepidoptera larvae only *Cadurciella tritaeniata, Xylotachina diluta* and *Townsendiellomyia nidicola* appear to be monophagous. Eggleton & Gaston (1992) analyse the generally high level of polyphagy found in the Tachinidae. The host ranges of tachinids attacking sawfly larvae (Hymenoptera: Symphyta) are also discussed by Pschorn-Walcher (1969).

Elsewhere in the Palaearctic other tachinids attack orthopterans and phasmids, and the genus *Strongygaster* attacks adult ants (see above). Outside the Palaearctic a few other groups of insects and other arthropods are also attacked by tachinids: a few genera attack mantids and cockroaches; the tribe Anacamptomyiini (Crosskey, 1976), the genus *Ophirion* and a species of *Lixophaga* (Wood, 1987) attack the immature stages of social wasps. Tachinids have also been reared from scorpions (Williams *et al.*, 1990), spiders (Vincent, 1985), Embioptera (Arnaud, 1963), adult tabanid flies (Spratt & Wolf, 1972) and adult calliphorid flies (Ferrar, 1977). Each of these groups is attacked by only one tachinid species.

Reproductive strategies

The Tachinidae fall into 6 groups according to the female reproductive system and method of host contact. This variety of strategies is one of the most interesting biological features of the Tachinidae. It contrasts strongly with the parasitoid Hymenoptera, which usually lay eggs directly on or in the host. The ancestral lack of a piercing ovipositor and the possession of more active larvae in the Tachinidae seem likely to be responsible for this difference.

(1) *Unincubated eggs laid on the host.* These eggs are ventrally flattened and stuck onto the host integument (see fig.306). The embryo develops on the outside of the host and, after 1–5 days (depending on the species, Herrebout, 1966: 350), the first-instar larva hatches and immediately penetrates the host integument. Hatching is either through an operculum (cap) at the anterior end of the egg or by boring through the floor of the egg directly into the host. A few Exoristiini (e.g. *Phorocera*) have modified eggs and female terminalia for partial insertion of the egg into the host integument (Herting, 1963; Wood, 1972). In Britain this strategy is found in some Phasiinae and (within the Exoristinae) the Exoristini, Winthemiini, a few Blondeliini, and in *Aplomya* in the Eryciini.

(2) *Unincubated eggs inserted into the host.* This group differs from group 1 in having an egg that is laid directly into the host through a wound. This is made using a piercing structure on the female terminalia separate from the ovipositor. Most of the British Phasiinae employ this strategy.

(3) *Incubated eggs laid on the host.* These eggs contain a fully developed larva and hatch very quickly, the first-instar larva immediately starting to penetrate the host's integument. Hatching and penetration together usually take less than 30 minutes and often only a few minutes (Herrebout, 1966: 350). In Britain this strategy is found in the following: some of the Blondeliini and most of the Eryciini

(Exoristinae); the Dufouriini and Voriini (Dexiinae); and some Tachininae. In the last group it is not known in all cases whether eggs are laid directly on the host or very close to it.

(4) *Incubated eggs inserted into the host.* This group differs from group 3 in that the egg is laid directly into the host through a wound. This is made using a piercing structure on the female terminalia separate from the ovipositor. In Britain this strategy is found only in *Blondelia nigripes, Compsilura concinnata* and *Vibrissina debilitata* (Exoristinae: Blondeliini).

(5) *Host contacted by the first-instar larva.* In this group an incubated egg is laid not on the host but in the host habitat. Host-contact is made by the newly-hatched tachinid larva. In Britain this group may be sub-divided into 3.
  (a) Certain Neaerini, Siphonini and Leskiini (Tachininae), together with *Pseudoperichaeta nigrolineata* and *Lydella* spp. (Exoristinae: Eryciini) lay eggs at the gallery entrance of stem-boring Lepidoptera larvae, or on the webs of leaf-rolling or other partially-concealed host species.
  (b) The majority of Tachininae have first-instar larvae which contact exposed leaf-feeding hosts, usually larval Lepidoptera. In some of the larger species (some *Ernestia, Eurithia* and *Linnaemya*) the larva is heavily sclerotised (presumably to prevent desiccation) and remains attached to the leaf at its posterior end via the remains of the egg capsule (see fig.307). The larva makes pendulum-type movements with its anterior end, seeking a passing host into which it bores. In other species the larva is mobile, moving freely over the leaf surface seeking a host, sometimes with the aid of ambulatory spines.
  (c) The Microphthalmini (Tachininae) and Dexiini (Dexiinae) have very active larvae which search for large Coleoptera larvae under bark and in the soil.

(6) *Eggs ingested by the host.* Species in this group lay very small eggs containing fully developed larvae. These heavily sclerotised 'microtype' eggs are laid on the host's food-plant. In contrast with group 5, they do not hatch until they are ingested by the host. The first-instar larva then penetrates the host's gut wall. Kahrer (1984) showed that in *Elodia morio* hatching is brought about by the combination of a lack of oxygen in the host gut, which induces movement of the tachinid larva, and the presence of a protease, which softens the wall of the egg. This strategy is found in (and defines) the Goniini (Exoristinae) (*sensu* Herting, 1984).

Strategies 5 and 6 allow tachinids to parasitise hosts which are inaccessible to the adult female, e.g. concealed hosts in plants such as leaf-rollers and stem-borers, hosts in the soil, and nocturnally-feeding hosts.
  In species where there is no appreciable uterine incubation (= ovipary), eggs only enter the vagina from the ovarioles after the preceding egg has been laid. The total egg production in these species is low, usually only between 100 and 200.
  In species which lay incubated eggs (= ovolarvipary), these descend continually from the ovarioles and accumulate in the vagina, which acquires the function of a uterus (see fig.305). In some species the uterus expands greatly as the eggs accumulate, eventually occupying most of the abdominal cavity. When the eggs are laid, hatching is sometimes so rapid that it is not clear whether in fact eggs or larvae have been laid. Wood (1987) states that egg cases are not normally found in the uterus of dissected females, but Andersen (pers. comm.) considers that the Siphonini (Tachininae) are in fact larviparous.
  In species which incubate their eggs there is a period between mating and the beginning of laying, e.g. 2–3 weeks in *Senometopia pollinosa* (Herrebout, 1969), 4

weeks in *Lypha dubia* (Schröder, 1969) and *Cyzenis albicans* (Hassell, 1968). *Elodia morio* can lay the first egg after 7 days and has its uterus full of incubated eggs after approximately 23 days (Kahrer, 1987). *Blondelia nigripes* begins to lay after 7–8 days (Herting, 1960).

In species which lay incubated eggs on or in the host (or in its immediate vicinity) egg production is usually similar to that of species which do not incubate. Among British species this ranges from approximately 50 incubated eggs carried in the uterus and ovariole bases of *Microsoma exigua* (Berry & Parker, 1950), to up to 600 eggs carried in the uterus of *Townsendiellomyia nidicola*.

In contrast, species which lay their eggs some distance from the host (i.e. where contact is made via a larva left on a leaf or searching through the soil, or the egg is ingested by the host) usually have higher egg production. Among British species *Ernestia rudis* carries approximately 1000 eggs in its uterus, *Dexia rustica* up to 600, *Cyzenis albicans* up to 2000 and *Zenillia libatrix* 800–2400. A North American *Gonia* is recorded as laying approximately 4000 eggs, and the European *Microphthalma europaea* of carrying that number (Herting, 1960). This higher egg production presumably compensates for the low probability that any individual egg or larva will contact a suitable host.

Only a few species appear to lay eggs which are partially incubated, e.g. *Zaira cinerea*. In the laboratory, however, young females of other species may lay incompletely incubated eggs.

Host location and selection

Adult females of several species of Tachinidae are attracted by olfactory stimuli. The odour of the host food-plant is attractive to female *Drino bohemica* (Monteith, 1955) while *Triarthria setipennis* adults are attracted to the odour of their host (Phillips, 1983). During their preoviposition period female *Senometopia pollinosa* are attracted to the odour of oak, but at the end of this period they are (temporarily) attracted to the odour of the host food-plant (pine) (Herrebout & Veer, 1969). In this species host location appears to be entirely visual, once the host-plant has been located, with the position of the host on the pine needle important (Herrebout, 1969). In North America *Hemyda aurata* and *Euclytia flava* have been caught in large numbers (both males and females) on sticky traps baited with a component of the host male pheromone (Aldrich *et al.*, 1984).

In North America *Ormia (=Euphasiopteryx) ochracea* (Ormiini) is attracted to the taped song of its cricket host (Cade, 1975).

The factors affecting oviposition vary with the method of host contact. In species where the host is contacted by the larva, or the egg is ingested, oviposition often takes place away from the host. Among species whose larvae search on leaves for the host, the presence of host silk (e.g. *Lypha dubia* (Cheng, 1969) and *Tachina magnicornis* (Townsend, 1908)), or host frass (e.g. *Linnaemya comta* (Clement *et al.*, 1986) and *Eurithia consobrina* (Herting, 1960)) can be sufficient. Clement *et al.* (1986) suggested that the ovipositional response of *L. comta* was mainly to a host metabolic product, and that the observed reduced response to plant damage was due to the presence of vomitus. Hsiao *et al.* (1966) and Nettles & Burks (1975) also suggested that oviposition was in response to a host metabolite. However, Thompson *et al.* (1983) found that extracts of host food-plant, as well as host frass, stimulated oviposition in *Lixophaga diatraeae*, and suggested the frass was merely concentrating plant chemicals. Among species laying eggs which are ingested by the host, leaf damage may be sufficient to elicit oviposition (e.g. in *Cyzenis albicans* (Hassell, 1968; Roland *et al.*, 1989)).

Dexia rustica, whose larvae search through the soil for hosts, appears to oviposit solely in response to a suitable host habitat (Walker, 1943).

In species where the host is contacted by the adult female, tarsal contact and visual cues are used. Oviposition of *Senometopia pollinosa* appears to follow tarsal contact but with little discrimination at this stage, some completely unsuitable hosts being readily accepted if the correct visual cues are given (Herrebout, 1969). In contrast, *Compsilura concinnata* does not appear to respond directly to hosts more than 5mm away, and host movement was only important following contact (when it helps stimulate attack). An agar-filled host integument presented alternately with live hosts elicited an equal ovipositional response (Weseloh, 1980).

Female tachinids do not appear to avoid ovipositing on or in hosts which already contain the immature stages of the same or other parasitoid species.

Oviposition and penetration of host integument

In species where the adult female oviposits on or in the host the action is usually very quick and does not allow the host to react. Female *Uromacquartia trinitatis* hold the arctiid larva (Lepidoptera) off the ground while they oviposit, lying on their back when host individuals are large (Cruttwell, 1969).

Species ovipositing on exposed leaf-feeding larvae usually place the eggs on the anterior part of the host, often on the first two thoracic segments. *Townsendiellomyia nidicola* uses its long ovipositor to place eggs between the thoracic legs. Unincubated eggs placed on posterior parts of the host body are often destroyed by the host with its mandibles. The gregarious larvae of *Agelastica alni* have even been observed destroying *Meigenia* eggs on their neighbours (Herting, 1960). Unincubated eggs left on the outside of hosts are often lost when the host moults. A number of species attack newly moulted hosts preferentially, or reject motionless hosts (many Lepidoptera larvae become motionless before moulting). Species of *Phorocera* make incisions in the host integument adjacent to the egg — presumably to anchor it to the new cuticle by scar tissue (Herting, 1963). These may all be adaptations to reduce the mortality among eggs resulting from moults of the host. One may speculate that uterine incubation in the Tachinidae has evolved as a response to the mortality suffered by eggs left exposed on the host for long periods.

Species attacking adult Coleoptera or Hemiptera often use the ovipositor to push the eggs between sclerites or under the wings (often standing on the host to achieve this, see fig.301). *Rondania dimidiata* (Meigen) inserts its ovipositor into the oesophagus of its adult curculionid host (Coleoptera) while the latter is feeding. These methods may facilitate the entry of larvae into these more heavily sclerotised hosts.

A number of different piercing structures of the female terminalia have evolved independently in the Tachinidae (see 'Species biology'). These all function to puncture the host integument and allow access for the ovipositor. The ovipositor itself is never adapted for piercing. In species laying eggs on the host, the ovipositor may be capable of considerable extension, e.g. Winthemiini (Exoristinae).

Except for species where the adult female has piercing structures on the terminalia, all tachinid larvae enter the host by cutting through the host integument. First-instar tachinid larvae lack functional mandibles — the anterior part of the cephalo-pharyngeal skeleton is extended forwards instead, forming a hook- or hatchet-shaped structure (the latter sometimes has a serrated anterior margin). This may be an adaptation for piercing the host integument. Entry of the tachinid larva into the host takes between 20 seconds and 1 hour depending on both the species of tachinid and host (Herrebout, 1966: 351).

Physiological interactions with host

Larvae which penetrate the host integument often remain with their posterior spiracles in contact with the outside air. In a typical host response to wounding, cells migrate to the affected area. This results in the formation of a sclerotised sheath called a *respiratory funnel*, which extends inwards from the edges of the wound (see fig.306). This funnel does not extend over the anterior end of the larva, which is therefore able both to feed and to respire. How the larva prevents complete encapsulation is not known, but it may be by purely mechanical means (e.g. by feeding or movement).

In some species the respiratory funnel may continue as a thin transparent sheath which temporarily encloses the anterior end of the larva (usually in tachinid species which overwinter as first-instar larvae). Movement and feeding do not appear to be hindered and the sheath is presumably permeable.

In other tachinid species the larvae do not remain at the point of entry in a respiratory funnel. In these species the larvae either migrate to specific tissues or remain free in the haemocoel (e.g. *Cyzenis albicans* enters the salivary glands; a North American *Gonia* enters the supra-oesophageal ganglion; *Senometopia pollinosa* usually initially enters either muscle bands or the hypodermis of the gut or body wall, often showing a complicated pattern of migration (Herrebout, 1969)). Larvae which hatch from eggs inserted into the host may also migrate to specific tissues, those of *Compsilura concinnata* and *Blondelia nigripes* penetrate the gut wall and spend their first instar within the peritrophic membrane. Most species whose larvae do not initially form respiratory funnels later penetrate the body wall or branches of the tracheal system (often using sclerotised spines on the posterior spiracular plates). Around this wound a respiratory funnel forms in the same manner as described previously.

Once a respiratory funnel is formed most tachinid larvae remain within it until the host dies, after which some species move about feeding in the corpse. Young tachinid larvae feed only on haemolymph. Usually after their final moult they produce digestive secretions which kill the host and liquefy its organs (Herting, 1960). Development is then rapid.

In Britain 47 species have been found to form respiratory funnels, and only *Loewia foeda* and *Leskia aurea* are known not to do so (Herting, 1960). In both species, however, the third-instar larva penetrates the body wall and feeds with its posterior spiracles in contact with the outside air after the host has been killed. The second-instar larva of *Loewia foeda* also has a particularly thin integument, presumably to facilitate gaseous exchange across it. Larvae of *Senometopia pollinosa* overwinter with their posterior spiracles in contact with the air-filled subalar cavity of the host pupa.

The formation of a respiratory funnel may serve both to satisfy the greater respiratory demands of older larvae (with their reduced surface area to volume ratio) and to avoid asphyxiation by the host's encapsulation response. The migration of many first-instar larvae from the haemolymph into tissues is probably to avoid encapsulation. Herrebout (1969) found that the proportion of larvae of *Senometopia pollinosa* encapsulated was higher in the haemocoel than within tissues. Larvae of a North American *Gonia* which do not reach the supra-oesophageal ganglion quickly enough, following their entry into the haemocoel from the gut, are encapsulated and killed (Strickland, 1930). It is possible that immunity from encapsulation is acquired during the stay in these tissues. However, the first-instar larvae of many species remain in the haemocoel and it is not known how these avoid encapsulation. The respiratory adaptations of tachinid larvae and their relationship to the endoparasitoid habit are discussed in Keilin (1944) and Salt (1968).

Species which overwinter as larvae within their host often go into diapause at the same time as their hosts. This is presumably triggered by the host's hormonal system. Some species however have an endogenous diapause, usually at the end of the first larval instar. This diapause is ended by a specific change in the host, e.g. its final larval moult (*Towsendiellomyia nidicola*); its pupation (*Zenillia libatrix, Cyzenis albicans* and *Phytomyptera nigrina*); or the start of wing histogenesis in the pupa (*Senometopia pollinosa*). Subsequent development of the tachinid and consumption of the host is rapid.

This endogenous diapause also allows species to attack young host individuals which are too small to support complete development of the tachinid. *Pseudoperichaeta nigrolineata* is able to attack all the host larval instars except the first, always emerging at the end of the host's final larval stage or from the pupa. In this species both larval moults appear to be influenced by the moults of the host (Ramadhane *et al.*, 1987). Plantevin *et al.* (1986) discuss other examples of hormonal influence by the host on tachinid larvae. Some tachinids do not synchronise their development with their hosts. These species usually only attack host larvae which are at least half-grown. They develop rapidly, e.g. *Winthemia quadripustulata* larvae are capable of developing to maturity in 4 days following penetration (Herting, 1960).

Systematics and general literature

One view of the systematic position of the Tachinidae within the Calyptratae is shown in fig.302. However, the evolutionary relationships, both within the Tachinidae and between it and the other oestroid families, are unclear and have been subject to much debate. An analysis of ribosomal RNA has indicated that the Tachinidae may not by monophyletic (Vossbrinck & Friedman, 1989), but this is not the view of most taxonomists. Rognes (1986) and McAlpine (1989) discuss the morphological characters supporting the monophyly of the Tachinidae and other oestroid families.

The classification followed here is that of Herting (1984), which places an emphasis upon the female reproductive system. He divides the family into four subfamilies; the Exoristinae, Tachininae, Dexiinae, and Phasiinae. Tschorsnig (1985) was unable to find characters in the male genitalia supporting the monophyly of Herting's Exoristinae and Tachininae (or indeed to separate them from other, non-tachinid, Oestroidea). Cantrell & Crosskey (1989) use an alternative classification, with a different arrangement of tribes in the Exoristinae (= Goniinae of other authors) and Tachininae, and a different arrangement of genera in the Goniini and Eryciini.

Dr H.-P. Tschorsnig and Dr B. Herting are currently preparing a key and host-list for European Tachinidae. The most recent key to European species (excluding the Phasiinae and part of the Dexiinae) is in Mesnil (1944–1975 and 1980). Wood (1987) provides a key to genera, and Arnaud (1978) a host-list, for North America.

This work supersedes that of Emden (1954), which deals also with the other oestroid families (i.e. the Calliphoridae, Sarcophagidae, Oestridae and Rhinophoridae). Here, a key to family level only is provided for these other groups. The most recent continental European publications on the other oestroid families are: Rognes (1991, Calliphoridae), Herting (1961, Rhinophoridae) and Pape (1987, Sarcophagidae).

The post-1950 literature on the immature stages of Tachinidae is reviewed in Ferrar (1987). Tschorsnig (1985) examines the comparative morphology of male terminalia in Palaearctic Tachinidae; Cantrell (1988) that of the male and female terminalia in Australian Tachinidae. The biology of the family is described in detail in Clausen (1940) and Herting (1960). The latter work includes summaries of the biology of all European species (where it is known). Belshaw (1992) investigates the habitat associations of tachinids at a locality in southern England using Malaise traps.

Identification

Preparing specimens

With experience, certain species can be identified by a cursory examination under a microscope or hand lens (or even with the naked eye) but most specimens will need to be pinned. This is best done as in fig.303. The pin securing the specimen to the stage should be as fine as possible (thus doing the minimum damage to the specimen) while the pin used for handling the specimen and stage should be much larger. A good material for the stage is Plastazote. Larger specimens (over 1cm) may be mounted using a single pin if this is preferred.

Collection data should be written on labels which are placed on the same pin as the specimen. Fig.303 shows the categories of data usually recorded (in this case from a reared specimen). Reared tachinids should be preserved with all parts of the puparium and the host remains. Gelatin capsules are useful for this purpose. Very large host remains may be glued to pieces of card (making sure the glue is both strong and water-soluble, e.g. Lepage's). When dealing with gregarious species, note on the label the number of tachinid puparia obtained from the host. If more than 1 individual is then mounted, reference should be made to this, e.g. by writing '3rd of 5 specimens from host'. This avoids possible duplication of rearing records. Brief habitat descriptions are very useful, e.g. 'beech woodland' or 'disturbed ground'. If adults do not emerge from puparia which have been kept indoors for about 3 weeks, they should be placed in a (non-airtight) container in a cool place over the winter and then transferred to an environment at room temperature.

Adult tachinids collected with a hand-net should be pinned as soon as possible before they harden − keeping them damp (but not wet) will provide more time if required. The body of most adult tachinids is partially covered by a layer of minute scales, which produces a grey colour that contrasts with the black or red colour of the integument. This grey colour (referred to as dusting or pollinosity by other authors) can be lost from specimens which have been immersed in liquid, causing problems when using this key.

In some species the male terminalia need to be examined in order to confirm identification. With experience these species can be recognised during mounting and their terminalia extended out from (but still left attached to) the abdomen using a bent pin. Softening dried specimens to do this is very difficult and it is best to remove the whole abdomen (push it gently from side to side, and up and down until it falls off). The abdomen is then softened by placing it in a solution of potassium hydroxide, either heating it for a few minutes or leaving to soak overnight. The terminalia can then be dissected out in alcohol (see fig.310 for a schematic view of the structures involved). After this, soak the abdomen in 100% alcohol to remove all the water and then glue it to a piece of card on the same pin as the specimen. A few specimens should be experimented on to find the correct degree of softening: if under-softened, structures will tear and bristles will break off; if over-softened, the abdominal sclerites will come away from each other and the dusting will be lost. Dissected male terminalia should be kept with the specimen in a microvial with a little glycerine. If the pin is pushed through the bung at a slight angle this will prevent the glycerine seeping out.

Use of the keys

The following notes are intended to help with the keys.

1. Most couplets in the keys have an accompanying marginal illustration. If there is no name below the illustration then it is of a generalised tachinid and merely shows the character referred to in the key (indicated by an arrow). Marginal illustrations with an accompanying name are accurate representations of the genus or species concerned. Figure numbers above 300 refer to illustrations at the back of the book. Figs 310–314 show the terminology used in the keys. This is from McAlpine *et al.* (1981) with minor changes for brevity, e.g. abdominal tergite 3 is referred to simply as abdomen 3. (Readers familiar with other terminologies should check these figures before using the keys.)

2. Many species come out in more than one place in the keys. The reader should therefore not assume that a mistake has been made if a specimen does not fully satisfy either alternative in a couplet. Taking either alternative should lead to the same name.

3. The keys will not work with badly damaged specimens (unless the reader is experienced).

4. Square brackets contain confirmatory characters.

5. Bristles can be distinguished from hairs by their greater size and conspicuous tapering. Their accidental loss can be discovered by finding their pores. Nomenclature of bristle orientation on the head is shown in fig. 308.

6. The grey colour often referred to in the keys is the dusting or pollinosity of other authors. It is a covering of microscopic scales on the surface of the insect. The structural colour is usually either black or some shade of red or yellow.

7. If it is necessary to sex a specimen, unless another character is given use the presence (females) or absence (males) of proclinate parafrontal bristles (see fig.309). If another character is given it is because either the males also have proclinate parafrontal bristles, or there is an easier way of distinguishing the sexes.

8. A statement such as '4–5mm in length' is an approximate measurement of the length of the tachinid from front to rear. Body lengths are only given when they provide a useful character at a particular place in the keys.

9. The best way to measure the width of an eye is as follows: (1) measure the maximum width of the head as seen directly from above; (2) subtract the width of the vertex (see fig. 311); (3) divide the result by 2.

Key to oestroid families

This key will distinguish British Tachinidae from all other Diptera. It will also allow Oestroidea (see fig. 302) to be identified to family (following Kloet & Hincks, 1977).

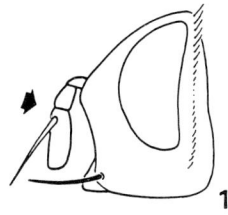

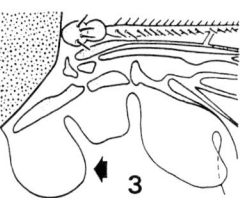

1   Antenna with 3 segments and a (hairlike) arista (fig.1) 2
–   Antenna with more than 3 segments ...... not in Oestroidea

2   Thorax with a suture (a linear indentation) extending unbroken across its upper surface (fig.2) – although this may become weaker in the middle 3
–   Thorax without such a suture or with it present only at the sides .... not in Oestroidea

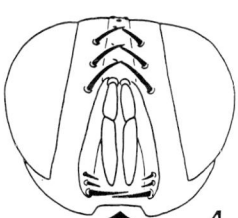

3   Base of the wing with a large flap (=lower calypter) next to the scutellum (fig.3) 4
–   Base of the wing without such a flap .... not in Oestroidea

4   Oral cavity possibly hidden by hairs and less than one tenth the width of the head. [10–14mm in length, body without bristles and possibly with a dense covering of hairs giving a beelike appearance] .... **Oestridae**
–   Oral cavity without any hairs covering it and at least one fifth the width of the head (fig.4) 5

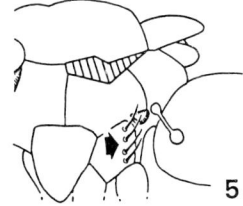

5   Thorax without meral (=hypopleural) bristles ....
.... in Muscoidea
–   Thorax with a row of at least 3 meral bristles (fig.5)......... 6

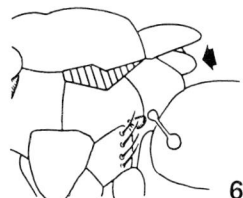

6   Subscutellum large and evenly rounded (fig.6) ....
.... **Tachinidae**
–   Subscutellum absent or not larger than in fig.338 7

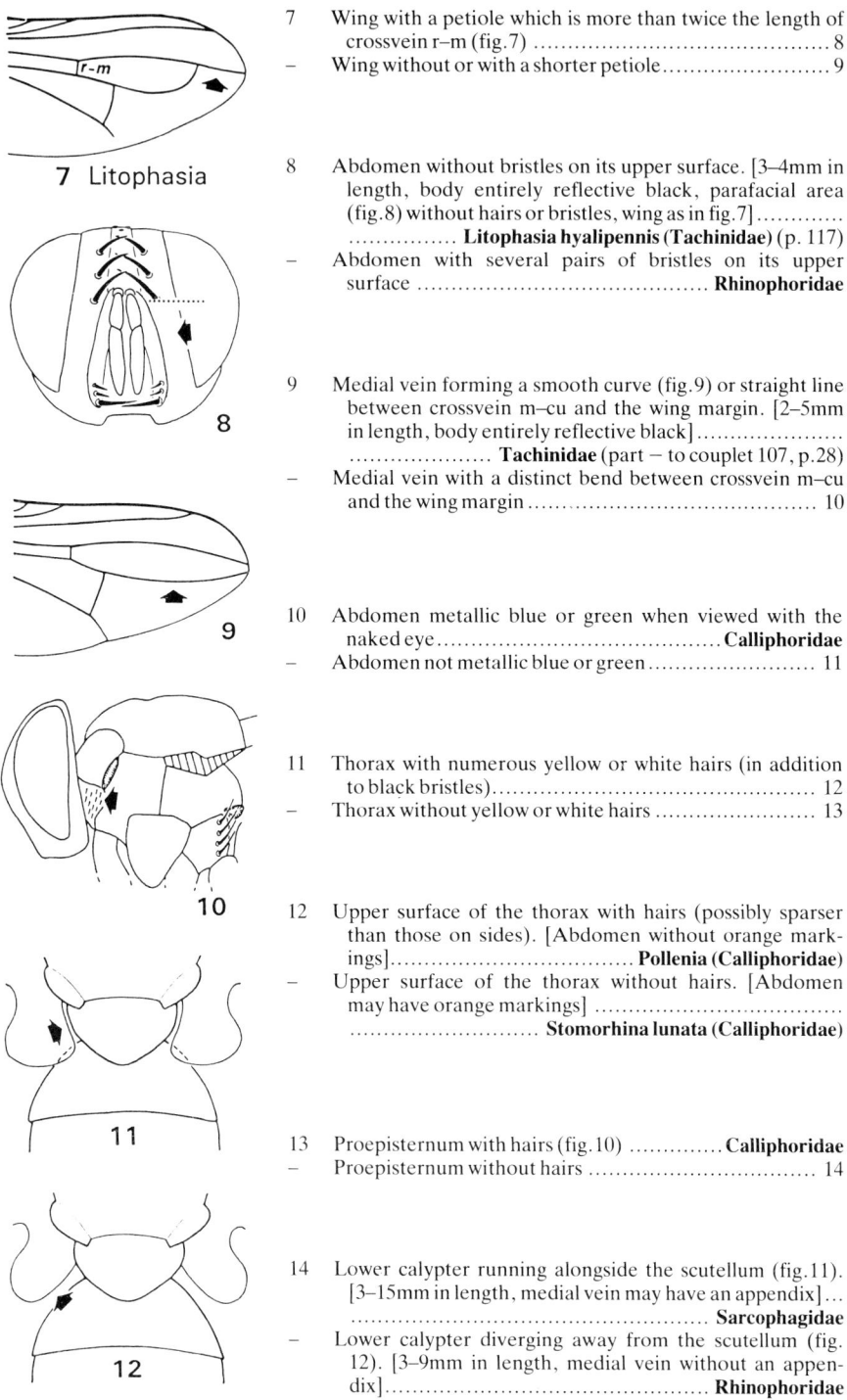

7 Litophasia

8

9

10

11

12

7   Wing with a petiole which is more than twice the length of crossvein r–m (fig.7) 8
–   Wing without or with a shorter petiole 9

8   Abdomen without bristles on its upper surface. [3–4mm in length, body entirely reflective black, parafacial area (fig.8) without hairs or bristles, wing as in fig.7] .... **Litophasia hyalipennis (Tachinidae)** (p. 117)
–   Abdomen with several pairs of bristles on its upper surface .... **Rhinophoridae**

9   Medial vein forming a smooth curve (fig.9) or straight line between crossvein m–cu and the wing margin. [2–5mm in length, body entirely reflective black] .... **Tachinidae** (part – to couplet 107, p.28)
–   Medial vein with a distinct bend between crossvein m–cu and the wing margin 10

10   Abdomen metallic blue or green when viewed with the naked eye ....**Calliphoridae**
–   Abdomen not metallic blue or green 11

11   Thorax with numerous yellow or white hairs (in addition to black bristles) 12
–   Thorax without yellow or white hairs 13

12   Upper surface of the thorax with hairs (possibly sparser than those on sides). [Abdomen without orange markings] .... **Pollenia (Calliphoridae)**
–   Upper surface of the thorax without hairs. [Abdomen may have orange markings] .... **Stomorhina lunata (Calliphoridae)**

13   Proepisternum with hairs (fig.10) ....**Calliphoridae**
–   Proepisternum without hairs 14

14   Lower calypter running alongside the scutellum (fig.11). [3–15mm in length, medial vein may have an appendix] ... . **Sarcophagidae**
–   Lower calypter diverging away from the scutellum (fig. 12). [3–9mm in length, medial vein without an appendix] .... **Rhinophoridae**

Key to genera

Excluding *Litophasia hyalipennis* (see key to families).
If specimen has been reared from a non-Lepidoptera
host also see table 1.
IMPORTANT. Please read the section explaining how
to use the key (p. 12). It will help avoid mistakes.

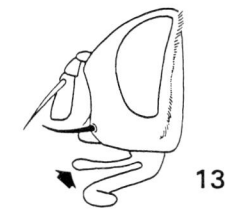

13

1　Very large fly (total length 1.5–2.0cm) with an entirely
　　black body and a yellow head ...... **Tachina grossa** (p. 87)
–　Not as above 2

2　Body with a bright metallic colour (usually green but
　　often partly bronze or blue and occasionally entirely
　　violet) 3
–　Body without a bright metallic colour (at the most a faint
　　dark-blue on the abdomen).................................... 4

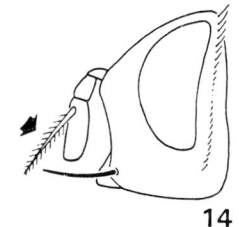

14

3　Palp (fig.13) dark-brown/black. [Scutellum with the most
　　apical pair of marginal bristles diverging and as long as
　　the adjacent pair] .... **Gymnocheta viridis** (p. 94)
–　Palp yellow. [Scutellum with the most apical pair of mar-
　　ginal bristles crossed (fig.47) and less than half the
　　length of the adjacent pair].....................................
　　...................................**Chrysocosmius auratus** (p. 90)

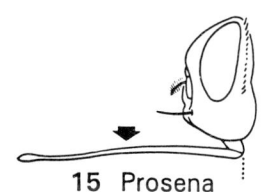

15 Prosena

4　Arista with hairs which are longer than its maximum dia-
　　meter (fig.14) 5
–　Arista without or with shorter hairs (view against a black
　　background) 15

5　Proboscis extremely elongated (fig.15) ....
　　...**Prosena siberita** (p.108)
–　Proboscis not longer than the height of the head 6

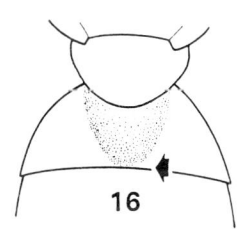

16

6　Face with a ridge between the antennae which is at least as
　　large as in fig.315... 7
–　Face without a ridge between the antennae (if the anten-
　　nae are touching then no ridge is present) 8

7　Excavation extending to the posterior margin of abdomen
　　1+2 (fig.16). [Abdomen may have large yellow mark-
　　ings (males), bend in the medial vein equal to a right
　　angle]... **Dexia** (p. 43)
–　Excavation not extending for more than two-thirds the
　　length of abdomen 1+2 (fig.17). [Abdomen without
　　yellow markings, bend in the medial vein more oblique]
　　...**Dinera** (p. 44)

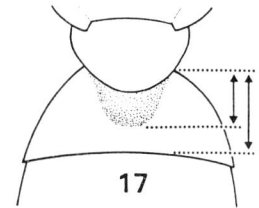

17

8　Wing with a petiole (fig.18) 9
–　Wing without a petiole (medial vein reaching the wing
　　margin independently of vein R4+5)...................... 10

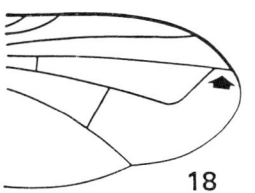

18

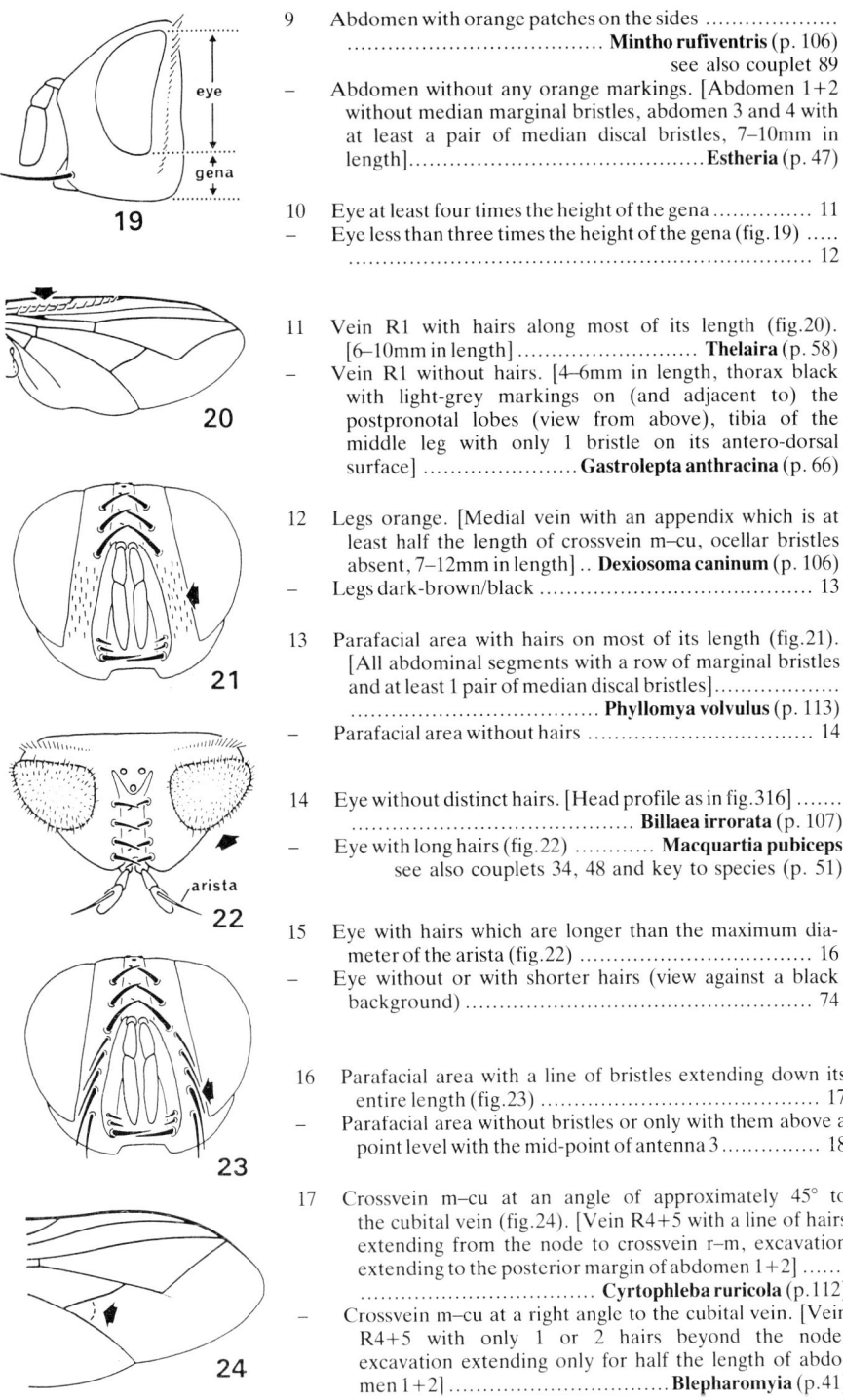

9   Abdomen with orange patches on the sides ....
.... **Mintho rufiventris** (p. 106)
see also couplet 89

–   Abdomen without any orange markings. [Abdomen 1+2 without median marginal bristles, abdomen 3 and 4 with at least a pair of median discal bristles, 7–10mm in length]...**Estheria** (p. 47)

10   Eye at least four times the height of the gena 11

–   Eye less than three times the height of the gena (fig.19) .....
.... 12

11   Vein R1 with hairs along most of its length (fig.20). [6–10mm in length] .... **Thelaira** (p. 58)

–   Vein R1 without hairs. [4–6mm in length, thorax black with light-grey markings on (and adjacent to) the postpronotal lobes (view from above), tibia of the middle leg with only 1 bristle on its antero-dorsal surface] ....**Gastrolepta anthracina** (p. 66)

12   Legs orange. [Medial vein with an appendix which is at least half the length of crossvein m–cu, ocellar bristles absent, 7–12mm in length] .. **Dexiosoma caninum** (p. 106)

–   Legs dark-brown/black 13

13   Parafacial area with hairs on most of its length (fig.21). [All abdominal segments with a row of marginal bristles and at least 1 pair of median discal bristles]..................
.... **Phyllomya volvulus** (p. 113)

–   Parafacial area without hairs 14

14   Eye without distinct hairs. [Head profile as in fig.316] ....
.... **Billaea irrorata** (p. 107)

–   Eye with long hairs (fig.22) .... **Macquartia pubiceps**
see also couplets 34, 48 and key to species (p. 51)

15   Eye with hairs which are longer than the maximum diameter of the arista (fig.22) 16

–   Eye without or with shorter hairs (view against a black background) 74

16   Parafacial area with a line of bristles extending down its entire length (fig.23) 17

–   Parafacial area without bristles or only with them above a point level with the mid-point of antenna 3 18

17   Crossvein m–cu at an angle of approximately 45° to the cubital vein (fig.24). [Vein R4+5 with a line of hairs extending from the node to crossvein r–m, excavation extending to the posterior margin of abdomen 1+2] ....
.... **Cyrtophleba ruricola** (p.112)

–   Crossvein m–cu at a right angle to the cubital vein. [Vein R4+5 with only 1 or 2 hairs beyond the node, excavation extending only for half the length of abdomen 1+2] ....**Blepharomyia** (p.41)

18   Facial ridge with a line of bristles extending up from the vibrissa at least half-way to the lower margin of the base of the antenna (fig.25) 19
–   Facial ridge with bristles or hairs not extending more than one-third of the way up 32

25

19   Edge of the mouth projecting forwards beyond the base of the vibrissa (fig.26) 20
–   Edge of the mouth not projecting forwards................. 21

26

20   Abdomen entirely black. [6–7mm in length, parafrontal area with at least 2 diverging bristles, excavation only extending for half the length of abdomen 1+2] ....
.... **Policheta unicolor** (p.67)
–   Abdomen with light-grey markings visible with the naked eye. [4–5mm in length, parafrontal area without diverging bristles, excavation extending to the posterior margin of abdomen 1+2].... **Chetogena acuminata** (p.62)
see also couplet 136

27

21   Scutellum with the most apical pair of marginal bristles more than 1.5 times the length of the scutellum (fig.27). [Ocellar bristles diverging and possibly slightly reclinate, parafrontal area without or (more rarely) with only a single reclinate bristle, 5–7mm in length] ....
.... **Campylocheta** (p.42)
–   Scutellum with the most apical pair of marginal bristles not more than 1.25 times the length of the scutellum. [Ocellar bristles at least slightly proclinate or absent].. 22

28

22   Tibia of the middle leg with only 1 large bristle on its antero-dorsal surface (fig.28) (any other bristles present are less than half its length) 23
–   Tibia of the middle leg with at least 2 large bristles on its antero-dorsal surface 25

23   Ocellar bristles absent. [Palp orange, scutellum without orange, female abdomen with teeth on its underside as in fig.326 (the piercer is not always visible)].................
.... **Compsilura concinnata** (p.69)
–   Ocellar bristles present (fig.29) 24

29

24   Scutellum with the most apical pair of bristles directed upwards from their base at an angle of approximately 45° to the horizontal (fig.30). [Thorax with 4 katepisternal bristles (occasionally 1 side with only 3), palp dark-brown/black, male terminalia as in fig.423] ....
.... **Pseudoperichaeta nigrolineata** (p.76)
–   Scutellum with the most apical pair of bristles not directed upwards. [Thorax with only 3 katepisternal bristles] ... 30

30

31

32

33

34
Diplostichus

35

36

25 Palp (fig.13) dark-brown/black 26
– Palp orange (possibly darker near its base) 27

26 Facial ridge with a line of large bristles (of similar size to most of the parafrontal bristles) which usually extend two-thirds of the way up the facial ridge (fig.25). [Tibia conspicuously lighter in colour than the body and other parts of the leg, parafrontal area with only 1 reclinate bristle]..................................... **Pales pavida** (p.82)
– Facial ridge with a line of bristles which decrease in size as they ascend (the upper ones only approximately one-quarter the length of the parafrontal bristles)............ 30

27 Scutellum with the basal pair of marginal bristles more than 2.25 times the length of the scutellum (fig.31) 28
– Scutellum with the basal pair of marginal bristles less than twice the length of the scutellum........................... 29

28 Abdomen 3 and 4 with a pair of median discal bristles (fig. 32). [Parafacial area with hairs on its upper part, thorax with 3 post-sutural dorso-central bristles] ....
....**Phorocera** (p.54)
– Abdomen 3 and 4 without median discal bristles. [Para-facial area without hairs (occasionally 1 or 2 hairs present just below the parafrontal bristles), thorax with 4 post-sutural dorso-central bristles].........................
...................................**Parasetigena silvestris** (p.63)

29 Scutellum without any bristles on its upper surface. [Bend in the medial vein forming a right angle, palp orange, thorax with the pre-alar bristle shorter than the 2 notopleural bristles (fig.34)]....................................
.... **Diplostichus janithrix** (p.63)
– Scutellum with a pair of bristles on its upper surface which (although small) are distinct from the surrounding hairs (fig.33) 30

30 Abdomen 1+2 without or with only a small pair of median marginal bristles (much thinner and only two-thirds the length of the median discal pair on abdomen 3) 31
– Abdomen 1+2 with a pair of large median marginal bristles (fig.35) 32

31 Parafrontal area with 2 reclinate bristles (fig.36). [4–6mm in length, apex of scutellum orange, male terminalia as in fig.424]............................. **Clemelis pullata** (p.82)
– Parafrontal area with at the most 1 reclinate bristle (some of the convergent bristles may also be slightly reclinate) 32

32   Abdomen and posterior half of thorax (= posterior to the suture) entirely reflective black (although they may have a faint red or green tinge when viewed under magnification).. 33

–   Abdomen and posterior half of thorax with light-grey markings visible with the naked eye 38

37

33   Scutellum with 3 pairs of marginal bristles (fig.37) (the most apical pair large and crossed). [The pair next to the most apical ones widely separated – the distance between their bases slightly greater than the length of the scutellum].. 34

–   Scutellum with 4 pairs of marginal bristles (the most apical pair may be very small). [The pair next to the apicals separated by less than the length of the scutellum]...... 35

38

34   Head profile similar to fig.318. [4–6mm in length, abdomen 1+2 usually with a row of marginal bristles]............ .... **Dufouria** (p.44)
see also couplets 90 and 126

–   Head profile similar to fig.319. [4–11mm in length, abdomen 1+2 without a row of marginal bristles] .... **Macquartia** (p.51)
see also couplet 48

39

35   Wing with a petiole which is at least as long as crossvein r–m (fig.38). [Vertex may be less than one-quarter the width of an eye viewed from above (males), 4–7mm in length]... **Loewia** (p.50)
see also couplet 101

–   Wing without a petiole (medial vein not joining vein R4+5 or doing so only at the wing margin). [Vertex at least two-thirds the width of an eye] 36

40

36   Gena more than 0.33 times the height of the eye (fig.39). [Thorax with 3 post-sutural dorso-central bristles] 37

–   Gena less than 0.30 times the height of an eye. [Thorax with 4 post-sutural dorso-central bristles (fig.41)] 38

41

37   Excavation distinct and reaching the posterior margin of abdomen 1+2 (fig.40). [3–7mm in length, thorax with 3 pre-sutural and 3 post-sutural acrostichal bristles, head profile as in fig.317]...................... **Lydina aenea** (p.90)

–   Excavation indistinct and not reaching the posterior margin of abdomen 1+2. [8–10mm in length, thorax with only the most posterior pair of acrostichal bristles present] .... **Zophomyia temula** (p.94)
see also couplet 124

38   Edge of the mouth projecting forwards beyond the base of the vibrissa (fig.42) 39

–   Edge of the mouth not projecting 43

42

43

44

45

46

47

48 Meigenia

39　Parafacial area without bristles (bristles not extending down from the parafrontal area beyond a point level with the tip of antenna 2) 40

－　Parafacial area with a line of bristles extending at least to a point level with one-third of the way down antenna 3 (fig.43) 55

40　Node of vein R4+5 with a hair that is at least twice the length of crossvein r–m (fig.44) 41

－　Node of vein R4+5 only with hairs that are not longer than crossvein r–m. [7–12mm in length] 42

41　Costal spine several times the length of the other bristles on the costa (fig.45). [5–9mm in length, head with inner vertical bristles crossed, proboscis quite long – mentum at least 0.7 times the height of the head] ....
.... **Eriothrix** (p.44)

－　Costal spine indistinct .... **Cadurciella tritaeniata** (p.77)
see also couplet 57

42　Wing with basicosta (fig.46) white/pale-yellow. [Palp absent or less than half the length of the antenna (therefore not visible if proboscis held within oral cavity), median vein with an appendix which is at least half as long as the distance between crossvein m–cu and the bend in the medial vein].............. **Linnaemya** (p.49)

－　Wing with basicosta dark-brown/black. [Palp as long as the antenna (therefore at least its tip always visible]........
.... **Ernestia** etc. (p.45)

43　Abdomen almost entirely orange (abdomen 3 and 4 either entirely orange (females) or with only an incomplete central dark stripe (males) which is less than one-tenth the width of abdomen 3). [5–7mm in length] ....
.... **Hyalurgus lucidus** (p.94)

－　Abdomen (excluding yellow or red patches on the sides) similar in colour to the thorax 44

44　Scutellum with the most apical pair of marginal bristles parallel or diverging 45

－　Scutellum with the most apical pair of marginal bristles crossed (fig.47) 48

45　Scutellum with the most apical pair of bristles only half the length and thickness of the adjacent pair – parallel and curved upwards (fig.48). [Male terminalia as in fig.432, note sexes are different colour].......................
.... **Meigenia majuscula** (p.65)

－　Scutellum with the most apical pair of bristles at least as large as the adjacent pair 46

46　Palp (fig.62) dark-brown/black. [Head with inner vertical bristles crossed, scutellum may have 4 pairs of (similarly sized) marginal bristles] .... **Cleonice callida** (p.94)

－　Palp yellow. [Head with inner vertical bristles not crossed, scutellum with 3 pairs of marginal bristles] 47

47 Legs entirely orange. [5–10mm in length, thorax with 4 post-sutural dorso-central bristles]..............................
.... **Phryno vetula** (p.83)
see also couplet 69
– Legs brown and grey. [4–6mm in length, parafrontal area in both sexes with at least 1 proclinate bristle, tibia of the middle leg with only a single bristle on its antero-dorsal surface].......................**Cyzenis albicans** (p.83)

49 Macquartia

48 Scutellum with 3 pairs of (similarly sized) marginal bristles (fig.49).......................................**Macquartia** (p.51)
see also couplet 34
– Scutellum with at least 4 pairs of marginal bristles (the most apical pair may be smaller than the adjacent pair) ...
.... 49

50

49 Parafacial area with hairs at least on its upper part (fig.50)
.... 50
– Parafacial area without or with only 1–3 hairs.............. 55

51

50 Thorax with only 2 katepisternal bristles (fig.51) (on both sides). [Abdomen 5 may have an orange band across its posterior margin] ....**Winthemia** (p.59)
– Thorax with 3 katepisternal bristles (occasionally 1 side with only 2). [Abdomen 5 without orange markings] .. 51

51 Arista shorter than antenna 3 and only tapering after at least four-fifths its length (fig.52). [Parafacial area very broad – minimum distance between the facial ridge and the eye 1.5 times the maximum width of antenna 3, 7–8mm in length].............. **Rhaphiochaeta breviseta** (p.70)
– Arista not as above. [Minimum width of parafacial area not greater than the maximum width of antenna 3] 52

52 Rhaphiochaeta

52 Upper part of the head with only white hairs behind the postocular row of hairs (fig.53). [Abdomen 3 and 4 without median discal bristles, tibia of the middle leg with more than 2 bristles on its antero-dorsal surface, male terminalia as in fig.437] ......**Timavia amoena** (p.70)
– Upper part of the head also with many black hairs behind the postocular row of hairs.................................... 53

53

53 Parafacial area with hairs over all or most of its length (fig.54) (extending at least as far as a point level with two- thirds of the way down antenna 3) 54
– Parafacial area with only half a dozen or so hairs (just below the parafrontal bristles) 60

54

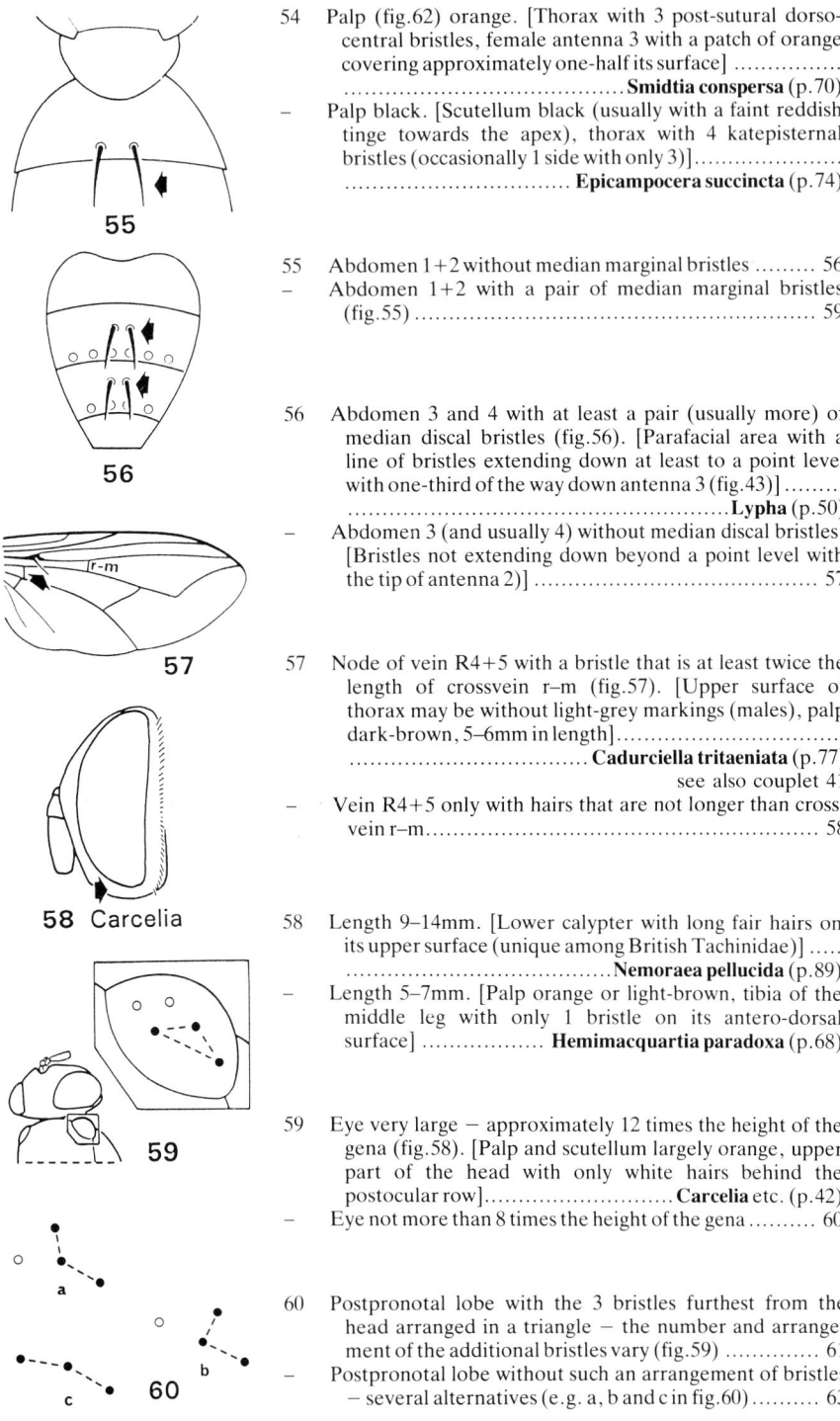

54   Palp (fig.62) orange. [Thorax with 3 post-sutural dorso-central bristles, female antenna 3 with a patch of orange covering approximately one-half its surface] ....
....**Smidtia conspersa** (p.70)

–   Palp black. [Scutellum black (usually with a faint reddish tinge towards the apex), thorax with 4 katepisternal bristles (occasionally 1 side with only 3)].....................
.... **Epicampocera succincta** (p.74)

55   Abdomen 1+2 without median marginal bristles 56
–   Abdomen 1+2 with a pair of median marginal bristles (fig.55) 59

56   Abdomen 3 and 4 with at least a pair (usually more) of median discal bristles (fig.56). [Parafacial area with a line of bristles extending down at least to a point level with one-third of the way down antenna 3 (fig.43)] ....
....**Lypha** (p.50)

–   Abdomen 3 (and usually 4) without median discal bristles. [Bristles not extending down beyond a point level with the tip of antenna 2)] 57

57   Node of vein R4+5 with a bristle that is at least twice the length of crossvein r–m (fig.57). [Upper surface of thorax may be without light-grey markings (males), palp dark-brown, 5–6mm in length]...................................
.... **Cadurciella tritaeniata** (p.77)
see also couplet 41

–   Vein R4+5 only with hairs that are not longer than cross-vein r–m... 58

58   Length 9–14mm. [Lower calypter with long fair hairs on its upper surface (unique among British Tachinidae)] .....
......................................**Nemoraea pellucida** (p.89)

–   Length 5–7mm. [Palp orange or light-brown, tibia of the middle leg with only 1 bristle on its antero-dorsal surface] .... **Hemimacquartia paradoxa** (p.68)

59   Eye very large – approximately 12 times the height of the gena (fig.58). [Palp and scutellum largely orange, upper part of the head with only white hairs behind the postocular row]............................**Carcelia** etc. (p.42)

–   Eye not more than 8 times the height of the gena 60

60   Postpronotal lobe with the 3 bristles furthest from the head arranged in a triangle – the number and arrange-ment of the additional bristles vary (fig.59) 61

–   Postpronotal lobe without such an arrangement of bristles – several alternatives (e.g. a, b and c in fig.60).......... 63

61 Upper part of the head with only white hairs behind the postocular row of hairs (fig.61). [Thorax with 2 katepisternal bristles, scutellum without orange] ....
....**Nemorilla floralis** (p.71)
– Upper part of the head also with many black hairs behind the postocular row of hairs. [Thorax with 3 katepisternal bristles] 62

61

62 Palp (fig.62) dark-brown. [Postpronotal lobe with at least 1 large bristle in addition to the 3 in the triangular arrangement (approximately the same size as those forming the triangle)] .... **Myxexoristops** (p.52)
– Palp orange (darker towards its base). [Postpronotal lobe with only smaller additional bristles (not more than two-thirds the length and thickness of those forming the triangle)]....................................... **Phebellia** (p.53)

62

63 Parafrontal area with only a single reclinate bristle (the convergent bristle nearest to the ocelli may also be partially reclinate but it is shorter and thinner) 64
– Parafrontal area with at least 2 reclinate bristles (fig.63) ....
.... 66

63

64 Palp (fig.62) with the apical part at least light-brown often orange, scutellum often orange at its apex. [Wing with vein R4+5 extending to the wing margin in a straight line and crossvein m–cu longer than the length of the cubital vein between it and the wing margin (fig.64), male terminalia as in fig.425] .... **Nilea hortulana** (p.73)
– Palp entirely dark-brown/black, scutellum without any orange... 65

R4+5

m–cu

64 Nilea

65 Anterior part of the thorax with a black stripe running down the middle when viewed from above and behind (fig.65), [Distance between the points where veins R2+3 and R4+5 meet the wing margin more than 1.5 times the distance between the latter and the wing tip (see fig.379), male terminalia as in fig.426] ....
.... **Eumea linearicornis** (p.81)
– Anterior part of the thorax without a black stripe running down the middle. [Distance between the points where veins R2+3 and R4+5 meet the wing margin less than 1.25 times the distance between the latter and the wing tip (see fig.380), male terminalia as in fig.427]..............
.... **Platymya fimbriata** (p.81)

65

66 Abdomen 3 without median discal bristles (occasionally some very small ones present)............................. 67
– Abdomen 3 with a pair of median discal bristles (fig.66) (of similar size to the median marginal pair) 71

66

67 Palp (fig.62) orange (possibly darker at its base) 68
– Palp dark-brown/black 70

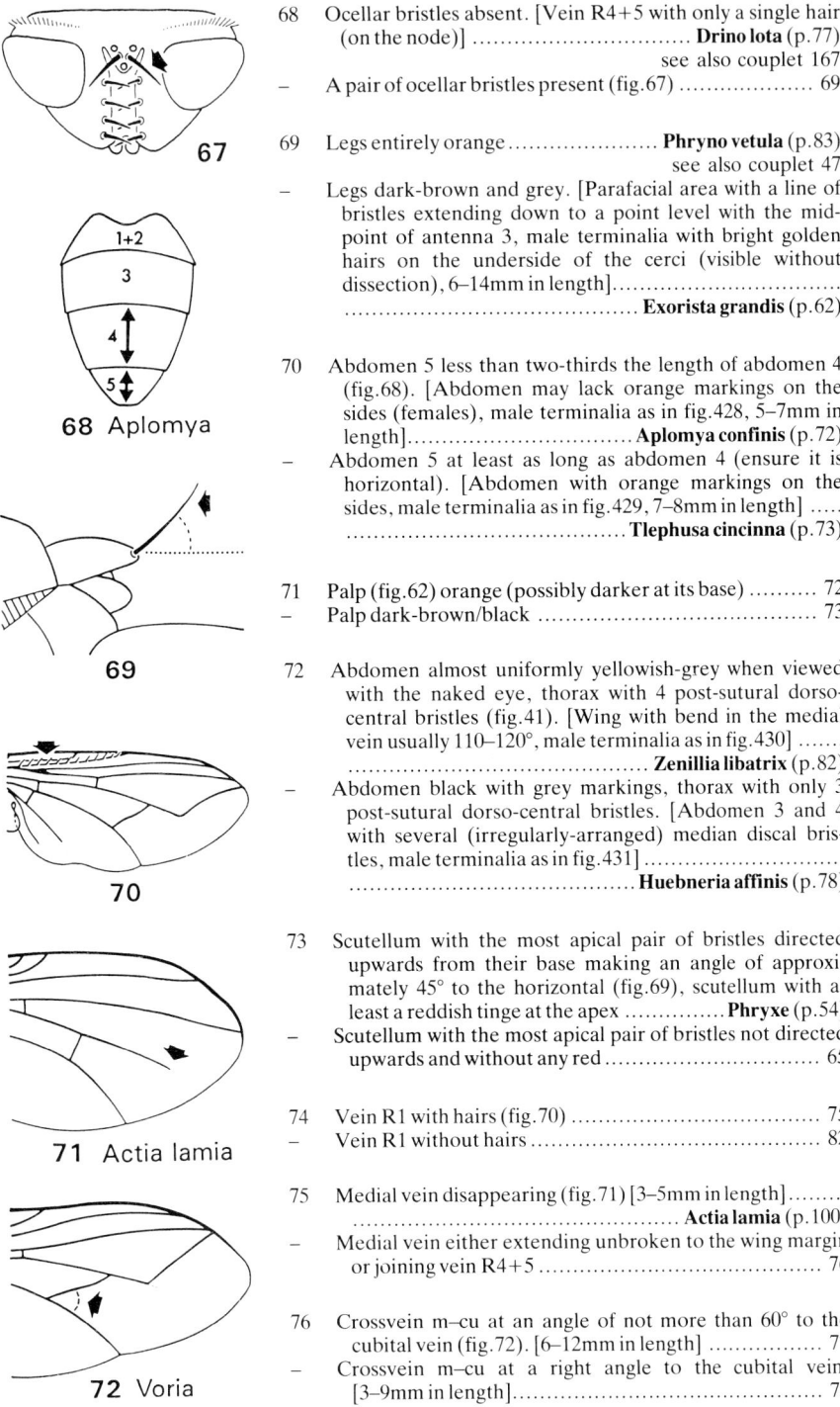

67

68 Aplomya

1+2
3
4
5

69

70

71 Actia lamia

72 Voria

68 Ocellar bristles absent. [Vein R4+5 with only a single hair (on the node)] .... **Drino lota** (p.77)
             see also couplet 167

– A pair of ocellar bristles present (fig.67) 69

69 Legs entirely orange .... **Phryno vetula** (p.83)
             see also couplet 47

– Legs dark-brown and grey. [Parafacial area with a line of bristles extending down to a point level with the mid-point of antenna 3, male terminalia with bright golden hairs on the underside of the cerci (visible without dissection), 6–14mm in length]................................
          .... **Exorista grandis** (p.62)

70 Abdomen 5 less than two-thirds the length of abdomen 4 (fig.68). [Abdomen may lack orange markings on the sides (females), male terminalia as in fig.428, 5–7mm in length]................................**Aplomya confinis** (p.72)

– Abdomen 5 at least as long as abdomen 4 (ensure it is horizontal). [Abdomen with orange markings on the sides, male terminalia as in fig.429, 7–8mm in length] .....
          ....**Tlephusa cincinna** (p.73)

71 Palp (fig.62) orange (possibly darker at its base) 72
– Palp dark-brown/black 73

72 Abdomen almost uniformly yellowish-grey when viewed with the naked eye, thorax with 4 post-sutural dorso-central bristles (fig.41). [Wing with bend in the medial vein usually 110–120°, male terminalia as in fig.430] ....
          .... **Zenillia libatrix** (p.82)

– Abdomen black with grey markings, thorax with only 3 post-sutural dorso-central bristles. [Abdomen 3 and 4 with several (irregularly-arranged) median discal bristles, male terminalia as in fig.431] ....
          ....**Huebneria affinis** (p.78)

73 Scutellum with the most apical pair of bristles directed upwards from their base making an angle of approximately 45° to the horizontal (fig.69), scutellum with at least a reddish tinge at the apex ....**Phryxe** (p.54)

– Scutellum with the most apical pair of bristles not directed upwards and without any red 65

74 Vein R1 with hairs (fig.70) 75
– Vein R1 without hairs 82

75 Medial vein disappearing (fig.71) [3–5mm in length].........
          .... **Actia lamia** (p.100)

– Medial vein either extending unbroken to the wing margin or joining vein R4+5 76

76 Crossvein m–cu at an angle of not more than 60° to the cubital vein (fig.72). [6–12mm in length] 77

– Crossvein m–cu at a right angle to the cubital vein. [3–9mm in length].. 78

78   Wing with a petiole (fig.73) 79
–   Wing without a petiole 80

79   Parafacial area with a line of bristles extending down its entire length (fig.74) .... **Ramonda prunaria** (p.110)
–   Parafacial area without bristles. [Vein R4+5 with only 1 or 2 hairs] .... **Rondania fasciata** (p.114)
see also couplet 100

80   Parafacial area with hairs on its entire length and bristles on the lower part (fig.75). [Arista with first (=basal) segment longer than the second] ....
.... **Triarthria setipennis** (p.97)
–   Parafacial area without hairs 81

73

74

81   Crossvein m–cu joining the medial vein at a point twice as far from crossvein r–m as from the bend (fig.76). [6–9mm in length, abdomen with (occasionally faint) red markings on the sides, ocellar bristles diverging (forming a 180° angle when viewed from above)]...................
.... **Bithia spreta** (p.105)
–   Crossvein m–cu at least as close to crossvein r–m as to the bend. [3–6mm in length, scutellum with the pair of marginal bristles adjacent to the (small) apical pair converging (fig.400)]......................... **Actia** etc. (p.38)
see also couplet 87

75

82   Vein R4+5 with a line of hairs extending most of the way from the node to crossvein r–m (fig.78) 83
–   Vein R4+5 with hairs not extending beyond half-way between the node and crossvein r–m 91

83   Parafacial area with a line of bristles extending down its entire length (fig.74). [Wing with a petiole] ....
.... **Wagneria** etc. (p.58)
–   Parafacial area without bristles or with them only on the upper part (not extending beyond a point level with the mid-point of antenna 3)...................................... 84

76

84   Excavation extending to the posterior margin of abdomen 1+2 (fig.77) (may be rather shallow and not immediately obvious)... 85
–   Excavation not extending for more than three-quarters the length of abdomen 1+2 86

77

85   Vein R4+5 with the line of hairs extending beyond crossvein r–m. [Parafacial area with a line of bristles extending down to a point level with the mid-point of antenna 3 (fig.43), crossvein m–cu at an acute angle to the cubital vein (fig.72)] .... **Athrycia** (p.41)
–   Vein R4+5 with hairs not extending beyond crossvein r–m.. 133

78

79

80 Siphona

81

82

83

84 Erynnia

86 Abdomen 3 and 4 without median discal bristles. [3–6mm in length, scutellum with the pair of marginal bristles adjacent to the (small) apical pair converging (fig.400), tibia of the middle leg with only 1 bristle on its antero-dorsal surface] 87
– Abdomen 3 and 4 with at least a pair of (possibly irregularly placed) median discal bristles (fig.79) 88

87 Proboscis extremely elongated (fig.80) (length from base of haustellum approximately equal to twice the height of the head) .... **Siphona** (p.55)
– Proboscis not longer than the height of the head ....
.... **Actia** etc. (p.38)
see also couplet 81

88 Femora (fig.81) largely orange 89
– Femora entirely dark-brown/black 90

89 Abdomen with large orange markings. [Abdomen compressed laterally with its maximum depth greater than its maximum width (unique among British Tachinidae), medial vein with an appendix, 6–9mm in length] ....
.... **Mintho rufiventris** (p.106)
see also couplet 9
– Abdomen without orange markings. [Tibia of the middle leg with at least 6 bristles on its antero-dorsal surface (fig.82), scutellum with the pair of marginal bristles adjacent to the (small) apical pair converging (fig.400), 5–6mm in length] .... **Goniocera versicolor** (p.99)

90 Thorax and abdomen entirely reflective black (possibly with a faint reddish tinge). [4–6mm in length] ....
.... **Dufouria** (p.44)
see also couplets 34 and 126
– Thorax and abdomen with light-grey markings visible with the naked eye. [Palp orange (possibly darker at its base), tibia of the middle leg with several bristles on its antero-dorsal surface, postpronotal lobe with the 3 bristles furthest from the head arranged in a triangle (fig.59), 6–9mm in length] .... **Belida angelicae** (p.64)

91 Medial vein disappearing (fig.83) 92
– Medial vein either extending unbroken to the wing margin or joining vein R4+5 93

92 Legs entirely dark-brown/black. [3–4mm in length, cross-vein m–cu absent (unique among British Tachinidae)] ....
.... **Phytomyptera nigrina** (p.98)
– Legs largely orange. [5–7mm in length] ....
.... **Ocytata pallipes** (p.84)
see also couplet 132

93 Wing with a petiole (fig.84) 94
– Wing without a petiole (medial vein not joining vein R4+5 or only doing so at the wing margin) 106

94 Abdomen very elongated – approximately 3 times as long as broad (fig.85). [Abdomen with orange patches on the sides] .... **Cylindromyia** (p.43)
– Abdomen not more than twice as long as broad 95

85 Cylindromyia

95 Abdomen without distinct bristles 96
– Abdomen with numerous large bristles (easily distinguishable from the surrounding hairs) 97

96 Vertex (fig.86) less than one-quarter the width of an eye viewed from above .... **Phasia** (p.53)
– Vertex at least half the width of an eye..........................
.... **Gymnosoma** etc.(p.48)

vertex · eye

86

97 Scutellum with the most apical pair of marginal bristles crossed (fig.87). [Vertex may be not more than half the width of an eye viewed from above (males)] 98
– Scutellum with the most apical pair of marginal bristles parallel or diverging. [Vertex at least three-quarters the width of an eye].. 102

98 Scutellum with the most apical pair of marginal bristles extending further backwards than the adjacent pair (fig.87). [2–4mm in length] 99
– Scutellum with the most apical pair of marginal bristles not extending as far back as the adjacent pair. [Abdomen entirely reflective black] 101

87

99 Wing with petiole not longer than the length of crossvein r–m (fig.88). [Female as in fig.301 (antenna 3 and palp orange in female but dark-brown/black in male)]...........
.... **Microsoma exigua** (p.114)
– Wing with petiole more than twice the length of crossvein r–m... 100

r–m

88

100 Abdomen with orange markings (fig.89). [Area separating the parafrontal areas largely orange, female abdomen as in fig.327 (antenna, palp and femora orange in female but dark-brown/black in male)] ....
.... **Rondania fasciata** (p.114)
see also couplet 79
– Abdomen without orange markings. [Face and legs dark-brown/grey, female abdomen without protruding terminalia] .... **Graphogaster brunnescens** (p.99)

89

101 Abdomen 3 and 4 with several median discal bristles (fig.90). [Excavation extending for approximately 0.9 times the length of abdomen 1+2 (which is usually without median marginal bristles)]........... **Loewia** (p.50)
see also couplet 35
– Abdomen 3 and 4 without median discal bristles. [Excavation not extending for more than half the length of abdomen 1+2 (which has a pair of median marginal bristles), female abdomen ending in a pair of forceps (similar to fig.344), 3–5mm in length] ....
.... **Leucostoma simplex** (p.118)

90

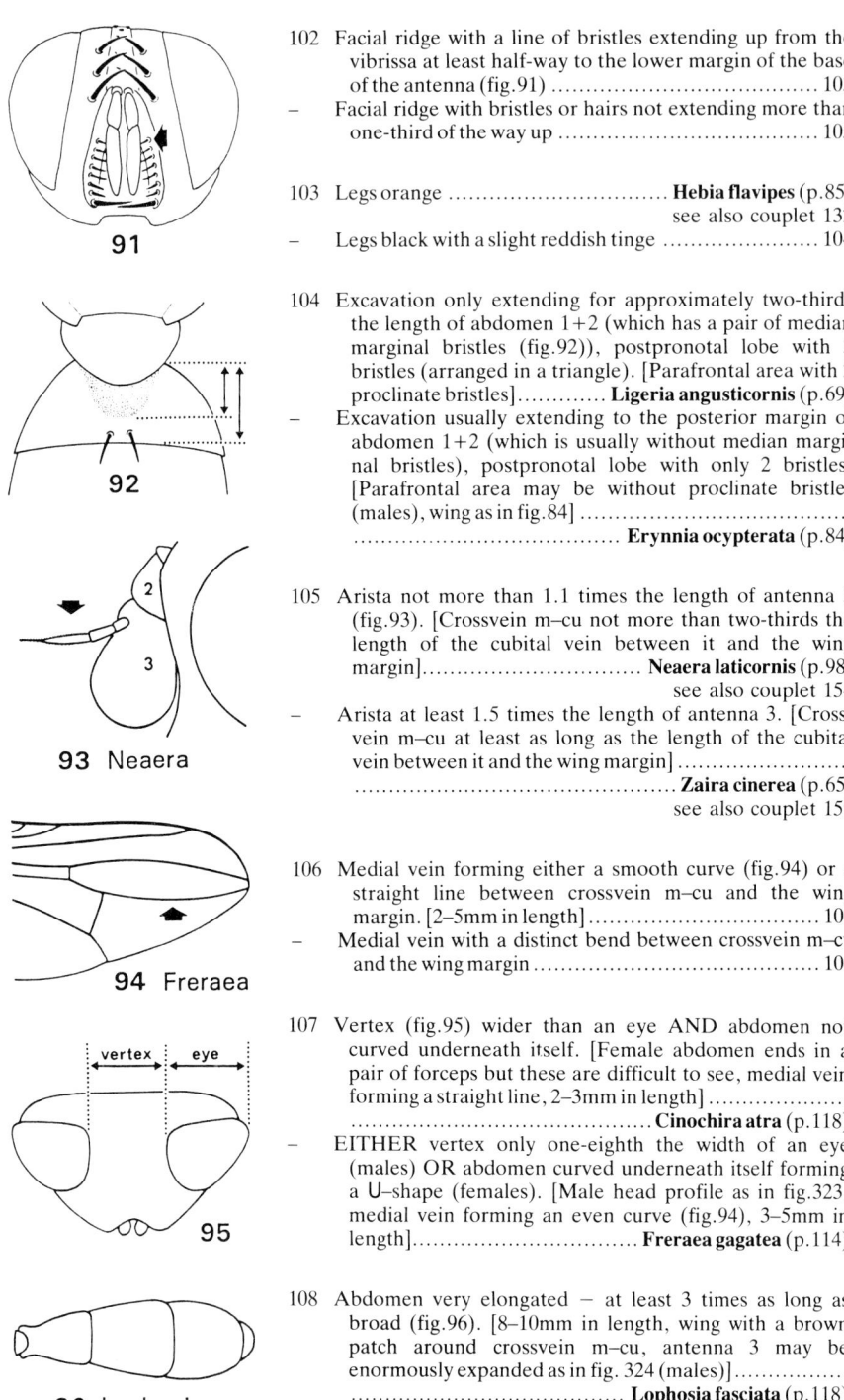

91

92

93 Neaera

94 Freraea

vertex · eye

95

96 Lophosia

102 Facial ridge with a line of bristles extending up from the vibrissa at least half-way to the lower margin of the base of the antenna (fig.91) 103
 – Facial ridge with bristles or hairs not extending more than one-third of the way up 105

103 Legs orange .... **Hebia flavipes** (p.85)
 see also couplet 132
 – Legs black with a slight reddish tinge 104

104 Excavation only extending for approximately two-thirds the length of abdomen 1+2 (which has a pair of median marginal bristles (fig.92)), postpronotal lobe with 3 bristles (arranged in a triangle). [Parafrontal area with 2 proclinate bristles]............. **Ligeria angusticornis** (p.69)
 – Excavation usually extending to the posterior margin of abdomen 1+2 (which is usually without median marginal bristles), postpronotal lobe with only 2 bristles. [Parafrontal area may be without proclinate bristles (males), wing as in fig.84] ....
 .... **Erynnia ocypterata** (p.84)

105 Arista not more than 1.1 times the length of antenna 3 (fig.93). [Crossvein m–cu not more than two-thirds the length of the cubital vein between it and the wing margin].............................. **Neaera laticornis** (p.98)
 see also couplet 154
 – Arista at least 1.5 times the length of antenna 3. [Crossvein m–cu at least as long as the length of the cubital vein between it and the wing margin] ....
 .... **Zaira cinerea** (p.65)
 see also couplet 159

106 Medial vein forming either a smooth curve (fig.94) or a straight line between crossvein m–cu and the wing margin. [2–5mm in length].................................. 107
 – Medial vein with a distinct bend between crossvein m–cu and the wing margin 108

107 Vertex (fig.95) wider than an eye AND abdomen not curved underneath itself. [Female abdomen ends in a pair of forceps but these are difficult to see, medial vein forming a straight line, 2–3mm in length] ....
 .... **Cinochira atra** (p.118)
 – EITHER vertex only one-eighth the width of an eye (males) OR abdomen curved underneath itself forming a U–shape (females). [Male head profile as in fig.323, medial vein forming an even curve (fig.94), 3–5mm in length]................................ **Freraea gagatea** (p.114)

108 Abdomen very elongated – at least 3 times as long as broad (fig.96). [8–10mm in length, wing with a brown patch around crossvein m–cu, antenna 3 may be enormously expanded as in fig. 324 (males)]................
 .... **Lophosia fasciata** (p.118)
 – Abdomen not more than twice as long as broad 109

109 Parafacial area very broad and with small bristles (fig.97). [9–13mm in length, head largely yellow, vertex twice the width of an eye viewed from above, ocellar bristles partially reclinate] .... **Gonia** (p.48)
– Parafacial area not as above 110

110 Body and femora with a dense fur-like covering of long yellow or yellow/white hairs. [9–14mm in length] .... **Tachina** subgenus **Servillia** (p.57)
– Body and femora without such hairs 111

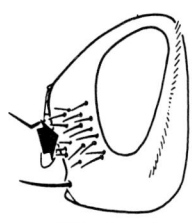

97 Gonia

111 Head with short orange antennae AND crossed inner vertical bristles (fig.98). [8–12mm in length, abdomen 1+2 with several (irregularly arranged) bristles adjacent to the median marginal pair].................... **Trixa** (p.58)
– Head not as above 112

112 Parafacial area with a line of bristles extending at least to a point level with one-third of the way down antenna 3 AND upper part of the head with only white hairs behind the postocular row of hairs (fig.99). [6–14mm in length].. **Exorista** (p.47)
– EITHER parafacial area without bristles OR upper part of the head also with black hairs behind the postocular row.. 113

98 Trixa

113 Abdomen with large orange markings..................... 114
– Abdomen without or with only small orange patches on the sides (narrower than the intervening dark stripe when viewed from above).................................. 121

114 Parafacial area with hairs extending down its entire length (fig.100) (these may be fair and therefore not immediately obvious). [Antenna 2 longer than antenna 3 (fig.101), 8–14mm in length] 115
– Parafacial area without hairs. [Antenna 2 shorter than antenna 3] 117

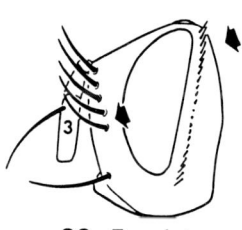

99 Exorista

115 Parafacial area with 2 or 3 bristles on its lower part (fig.101). [Ocellar bristles absent, palp narrow – its width similar to that of the arista]............................. .... **Peleteria rubescens** (p.88)
– Parafacial area without bristles. [Ocellar bristles present] 116

116 Antenna 2 and tibiae orange. [Palp narrow – its width similar to that of the arista].............. **Tachina fera** (p.87)
– Antenna 2 and tibiae dark-brown/black. [Palp with its maximum width similar to that of antenna 2] .... **Nowickia ferox** (p.88)

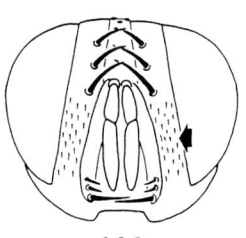

100

117 Abdomen almost entirely orange........................... 118
– Abdomen with a continuous grey or black stripe running down the middle (occasionally a narrow break at the junction of abdomen 1+2 with 3)........................... 119

101 Peleteria

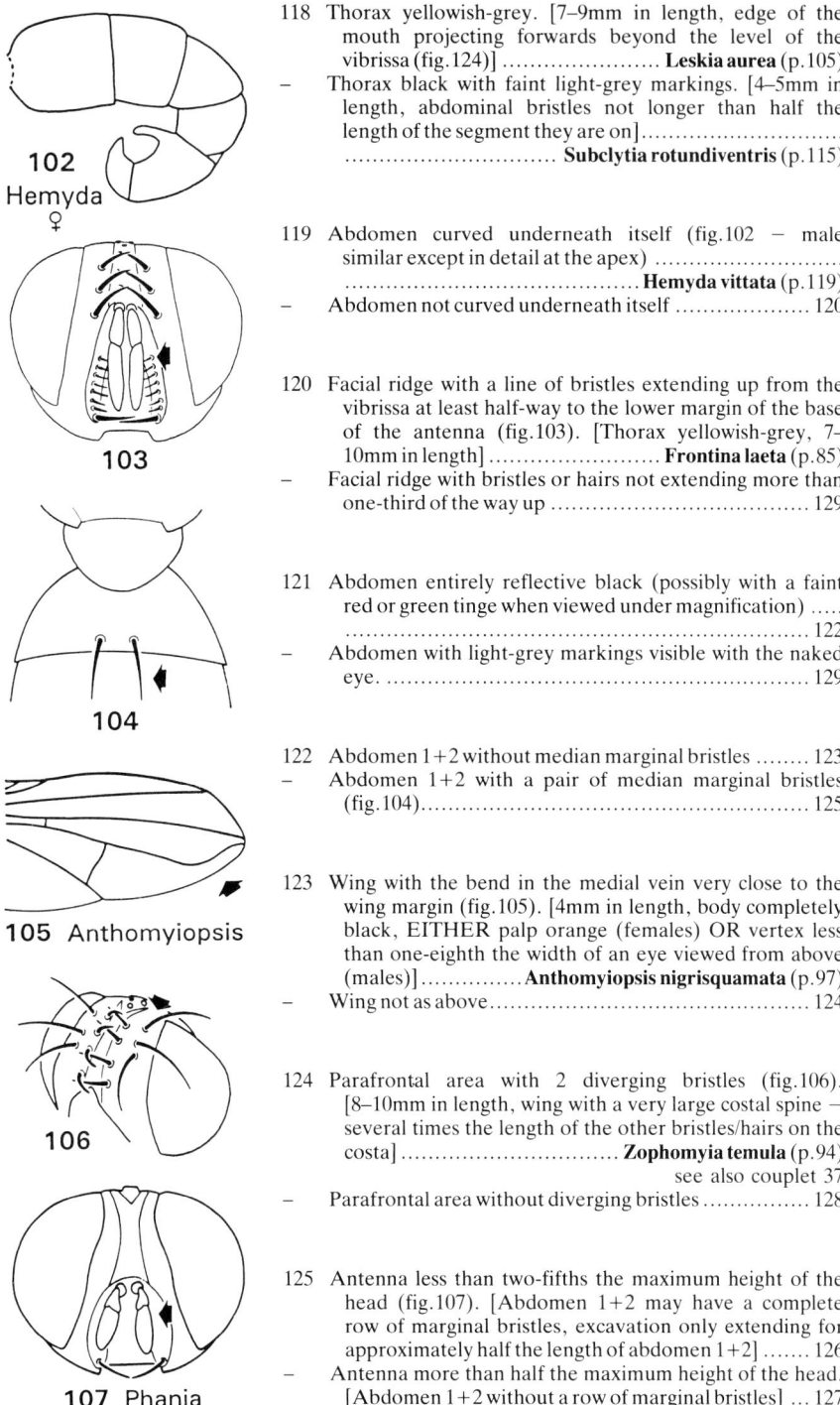

102
Hemyda
♀

103

104

105 Anthomyiopsis

106

107 Phania

118 Thorax yellowish-grey. [7–9mm in length, edge of the mouth projecting forwards beyond the level of the vibrissa (fig.124)] .... **Leskia aurea** (p.105)
– Thorax black with faint light-grey markings. [4–5mm in length, abdominal bristles not longer than half the length of the segment they are on]............................ .... **Subclytia rotundiventris** (p.115)

119 Abdomen curved underneath itself (fig.102 – male similar except in detail at the apex) ....**Hemyda vittata** (p.119)
– Abdomen not curved underneath itself 120

120 Facial ridge with a line of bristles extending up from the vibrissa at least half-way to the lower margin of the base of the antenna (fig.103). [Thorax yellowish-grey, 7–10mm in length] .... **Frontina laeta** (p.85)
– Facial ridge with bristles or hairs not extending more than one-third of the way up 129

121 Abdomen entirely reflective black (possibly with a faint red or green tinge when viewed under magnification) 122
– Abdomen with light-grey markings visible with the naked eye. 129

122 Abdomen 1+2 without median marginal bristles 123
– Abdomen 1+2 with a pair of median marginal bristles (fig.104)... 125

123 Wing with the bend in the medial vein very close to the wing margin (fig.105). [4mm in length, body completely black, EITHER palp orange (females) OR vertex less than one-eighth the width of an eye viewed from above (males)] .... **Anthomyiopsis nigrisquamata** (p.97)
– Wing not as above... 124

124 Parafrontal area with 2 diverging bristles (fig.106). [8–10mm in length, wing with a very large costal spine – several times the length of the other bristles/hairs on the costa] .... **Zophomyia temula** (p.94)
see also couplet 37
– Parafrontal area without diverging bristles 128

125 Antenna less than two-fifths the maximum height of the head (fig.107). [Abdomen 1+2 may have a complete row of marginal bristles, excavation only extending for approximately half the length of abdomen 1+2] 126
– Antenna more than half the maximum height of the head. [Abdomen 1+2 without a row of marginal bristles] ... 127

126 Abdomen 3 and 4 with discal bristles (fig.108 – number and arrangement vary). [Female abdomen not curved underneath itself] .... **Dufouria** (p.44)
see also couplets 34 and 90
– Abdomen 3 and 4 without discal bristles. [Female abdomen curved underneath itself (fig.328), head as in fig.107] .... **Phania** (p.52)

127 Vertex (fig.109) less than three-quarters (females) or less than half (males) the width of an eye. [3–4mm in length] .... **Medina** (p.52)
see also couplets 138 and 161
– Vertex as wide as an eye 128

128 Abdomen and thorax entirely black except sometimes for a faint grey on the postpronotal lobes. [Facial ridge with bristles extending up from the vibrissa approximately half-way to the lower margin of the base of the antenna (fig.110), 3–5mm in length] .... **Elodia morio** (p.85)
– Abdomen (and possibly also thorax) with light-grey markings when viewed under magnification 133

129 Femora (fig.111) entirely or largely orange 130
– Femora with at least half their area dark-brown or grey (similar to the body colour) 133

130 Abdomen 3 and 4 without median discal bristles. [Abdomen may have large orange markings (males)] ....
.... **Solieria** (p.57)
see also couplet 149
– Abdomen 3 and 4 with a pair of median discal bristles (fig.108). [Abdomen without orange markings, scutellum with at least 3 pairs of marginal bristles] 131

131 Abdomen 1+2 without median marginal bristles. [7–8mm in length, vertex may be only approximately one-tenth the width of an eye viewed from above (males), parafrontal area without reclinate bristles] ....
.... **Redtenbacheria insignis** (p.115)
– Abdomen 1+2 with a pair of median marginal bristles (fig.112). [Vertex wider than an eye] 132

132 Scutellum with the most apical pair of marginal bristles the longest (fig.113). [Antenna 3 may be yellow (females), tibia of the middle leg with only 1 long bristle on its antero-dorsal surface (any others not more than half its length), 4–6mm in length] .... **Hebia flavipes** (p.85)
see also couplet 103
– Scutellum with the most apical pair of marginal bristles only approximately half the length of the adjacent pair. [Antenna 3 dark-grey, tibia of the middle leg with 2 long bristles on its antero-dorsal surface] ....
.... **Ocytata pallipes** (p.84)
see also couplet 92

31

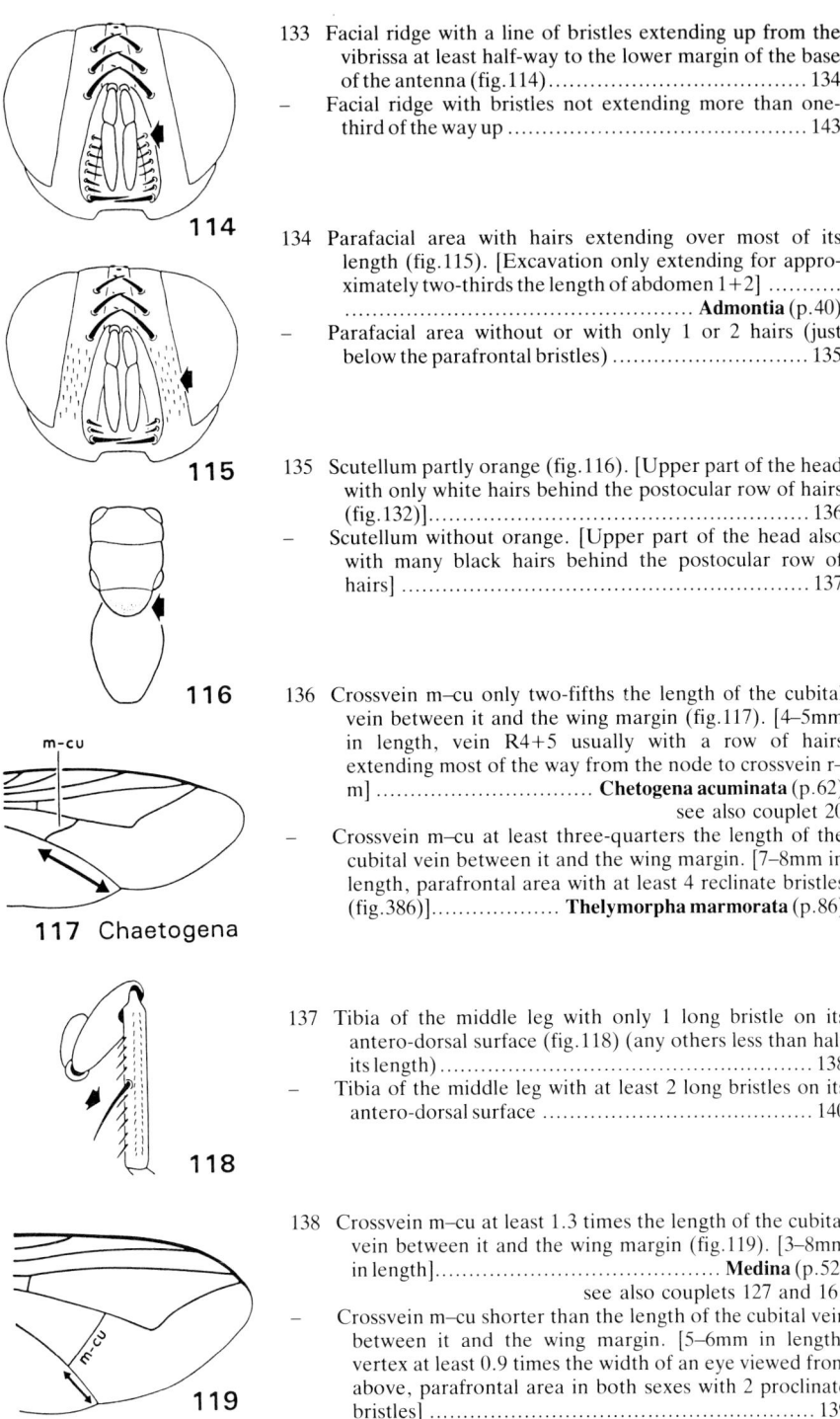

133 Facial ridge with a line of bristles extending up from the vibrissa at least half-way to the lower margin of the base of the antenna (fig.114) 134

– Facial ridge with bristles not extending more than one-third of the way up 143

114

134 Parafacial area with hairs extending over most of its length (fig.115). [Excavation only extending for approximately two-thirds the length of abdomen 1+2] .... **Admontia** (p.40)

– Parafacial area without or with only 1 or 2 hairs (just below the parafrontal bristles) 135

115

135 Scutellum partly orange (fig.116). [Upper part of the head with only white hairs behind the postocular row of hairs (fig.132)] 136

– Scutellum without orange. [Upper part of the head also with many black hairs behind the postocular row of hairs] 137

116

m–cu

136 Crossvein m–cu only two-fifths the length of the cubital vein between it and the wing margin (fig.117). [4–5mm in length, vein R4+5 usually with a row of hairs extending most of the way from the node to crossvein r–m] .... **Chetogena acuminata** (p.62) see also couplet 20

– Crossvein m–cu at least three-quarters the length of the cubital vein between it and the wing margin. [7–8mm in length, parafrontal area with at least 4 reclinate bristles (fig.386)] .... **Thelymorpha marmorata** (p.86)

117 Chaetogena

137 Tibia of the middle leg with only 1 long bristle on its antero-dorsal surface (fig.118) (any others less than half its length) 138

– Tibia of the middle leg with at least 2 long bristles on its antero-dorsal surface 140

118

138 Crossvein m–cu at least 1.3 times the length of the cubital vein between it and the wing margin (fig.119). [3–8mm in length] .... **Medina** (p.52) see also couplets 127 and 161

– Crossvein m–cu shorter than the length of the cubital vein between it and the wing margin. [5–6mm in length, vertex at least 0.9 times the width of an eye viewed from above, parafrontal area in both sexes with 2 proclinate bristles] 139

m–cu

119

139 Palp (fig.120) orange. [Excavation not reaching the posterior margin of abdomen 1+2 (a gap approximately
equal to one-fifth the length of the segment), female
abdomen with a series of short bristles along its
underside (fig.329 − distinguish sexes using terminalia
(fig.309))] .... **Vibrissina debilitata** (p.70)
see also couplet 160
– Palp dark-brown/black. [Excavation reaching the posterior margin of abdomen 1+2, thorax with 4 pairs of
post-sutural dorsal-central bristles] ....
.... **Elodia ambulatoria** (p.85)

120

140 Vertex (fig.121) twice the width of an eye. [Bend in the
medial vein closer to crossvein m–cu than to the nearest
point on the wing margin (fig.381)] ....
.... **Brachicheta strigata** (p.86)
– Vertex not more than 1.3 times the width of an eye.
[Bend in the medial vein closer to the wing margin than
to crossvein m–cu] 141

121

141 Crossvein m–cu longer than the length of the cubital vein
between it and the wing margin (fig.122). [Excavation
not extending for more than three-quarters the length of
abdomen 1+2 (usually only for half its length)] ....
.... **Oswaldia muscaria** (p.68)
see also couplets 160 and 164
– Crossvein m–cu not more than three-quarters the length
of the cubital vein between it and the wing margin.
[Excavation reaching the posterior margin of abdomen
1+2] 142

122

142 Vein R4+5 with at least 4 hairs (usually a line extending
most of the way from the node to crossvein r–m
(fig.123)). [3–5mm in length, scutellum with the most
apical pair of marginal bristles arranged parallel to each
other or slightly diverging (longer than the adjacent
pair)] .... **Bessa** (p.41)
– Vein R4+5 with only 1 or (more rarely) 2 hairs (these on
the node). [5–8mm in length, scutellum with the most
apical pair of marginal bristles crossed (only approximately half the length of the adjacent pair)] ....
.... **Lydella** (p.50)
see also couplet 171

123

143 Edge of the mouth projecting forwards beyond the base of
the vibrissa (fig.124).. 144
– Edge of the mouth not obviously projecting............... 151

124 Demoticus

144 Proboscis elongated − the haustellum at least 1.1 times
the height of the head and extending forwards beyond
the edge of the mouth (fig.125). [Vein R4+5 with a line
of hairs extending at least half-way from the node to
crossvein r–m (usually further)] ....
.... **Aphria longirostris** (p.104)
– Proboscis with the haustellum not more than three-
quarters the height of the head (not extending forwards
beyond the edge of the mouth). [Vein R4+5 with a
maximum of 2 hairs beyond the node]................... 145

125 Aphria

126

127

vertex

ocelli

128

129

130

1+2

3

131

145   Excavation reaching or almost reaching the posterior margin of abdomen 1+2 (fig.126) 146
 –   Excavation only extending for approximately half the length of abdomen 1+2 148

146   Abdomen 3 without or with very small median discal bristles (only half the length and thickness of the median marginal pair). [Thorax with only 3 post-sutural dorso-central bristles] .... **Erycia furibunda** (p.80)
                         see also couplet 170
 –   Abdomen 3 with a pair of large median discal bristles (fig.127). [Thorax with 4 post-sutural dorso-central bristles] 147

147   Scutellum (at least faintly) orange towards its apex (fig.116). [Parafrontal area may be without proclinate bristles (males), arista starts to taper after less than one-third its length] .... **Bithia modesta** (p.104)
 –   Scutellum without orange. [Parafrontal area with 2 pro-clinate bristles, arista starts to taper only after at least two-fifths its length] .... **Demoticus plebejus** (p.104)

148   Vertex very narrow – less than the distance between the ocelli (fig.128). [Head profile as in fig.325]...................
         .... **Opesia cana** (p.116)
 –   Vertex more than half the width of an eye 149

149   Scutellum with the most apical pair of marginal bristles arranged parallel to each other or diverging (fig.129) (usually only 2 pairs). [Abdomen 1+2 with only a pair of median marginal bristles] .... **Solieria** (p.57)
                         see also couplet 130
 –   Scutellum with the most apical pair of marginal bristles crossed (3 pairs present). [Abdomen 1+2 usually with some additional marginal bristles adjacent to the median pair, female abdomen with a pair of forceps at the end (similar to fig.344)]..................................... 150

150   Thorax with a pair of pre-sutural acrostichal bristles (fig.130), ocellar bristles diverging and slightly reclinate. [Postpronotal lobe with the bristles not all arranged in a line]............................... **Dionaea aurifrons** (p.117)
 –   Thorax without pre-sutural acrostichal bristles, ocellar bristles proclinate. [Postpronotal lobe with bristles in a line (occasionally only 2 present)].............................
          .... **Labigastera forcipata** (p.118)

151   Excavation not extending to the posterior margin of abdomen 1+2 (a gap equal to more than one-eighth the length of the segment) AND abdomen 3 without median discal bristles. [3–6mm in length, abdomen 1+2 usually without median marginal bristles]........................ 152
 –   EITHER excavation extending to the posterior margin of abdomen 1+2 OR abdomen 3 with a pair of median discal bristles (fig.131) 155

152 Head with only black hairs behind the postocular
    row of hairs (fig.132) 153
  – Head also with white hairs behind the postocular row
    (although ones nearer the postocular row may all be
    black)... 155

153 Scutellum with the most apical pair of marginal bristles
    crossed near their mid-point (fig.133) (longer than the
    adjacent pair). [Abdomen may have orange markings
    (males), head with the inner vertical bristles crossed,
    palp absent] ....**Eloceria delecta** (p.95)
  – Scutellum with the most apical pair of marginal bristles
    not crossed (occasionally meeting at their tips). [Abdo-
    men without orange markings, head with the inner
    vertical bristles not crossed, palp present but may be
    small] 154

154 Scutellum with the most apical pair of marginal bristles as
    long as the others (and without hairs between them). [5–
    6mm in length, excavation extending for at least four-
    fifths the length of abdomen 1+2]............................
    .... **Neaera laticornis** (p.98)
                                see also couplet 105
  – Scutellum with the most apical pair of marginal bristles
    not more than half the length of the adjacent pair
    (fig.134) (possibly difficult to distinguish from hairs).
    [3–4mm in length, excavation only extending for half the
    length of abdomen 1+2, head with a pair of bristles
    immediately posterior to the ocelli (similar to fig.280)]....
    .... **Phytomyptera cingulata** (p.98)

155 Scutellum with the most apical pair of marginal bristles
    arranged parallel to each other and curved upwards –
    approximately half the thickness and length of the
    adjacent pair (fig.135). [Thorax with 3 post-sutural
    dorso-central bristles, 3–8mm in length, N.B. sexes
    differ in colour]................................. **Meigenia** (p.52)
  – Scutellum with the most apical pair of marginal bristles
    not as above 156

156 Scutellum with the most apical pair of marginal bristles
    arranged parallel to each other or diverging – at least as
    long as the adjacent pair (fig.136) 157
  – Scutellum with the most apical pair of marginal bristles
    crossed – smaller than the adjacent pair. [Scutellum
    with more than 3 pairs of marginal bristles, thorax with 4
    pairs of post-sutural dorso-central bristles] 165

157 Palp (fig.152) orange 158
  – Palp dark-brown or black 161

158 Abdomen 3 without median discal bristles 159
  – Abdomen 3 with a pair of median discal bristles (fig.137)
    160

132 P. cingulata

133

134

135 Meigenia

136

137

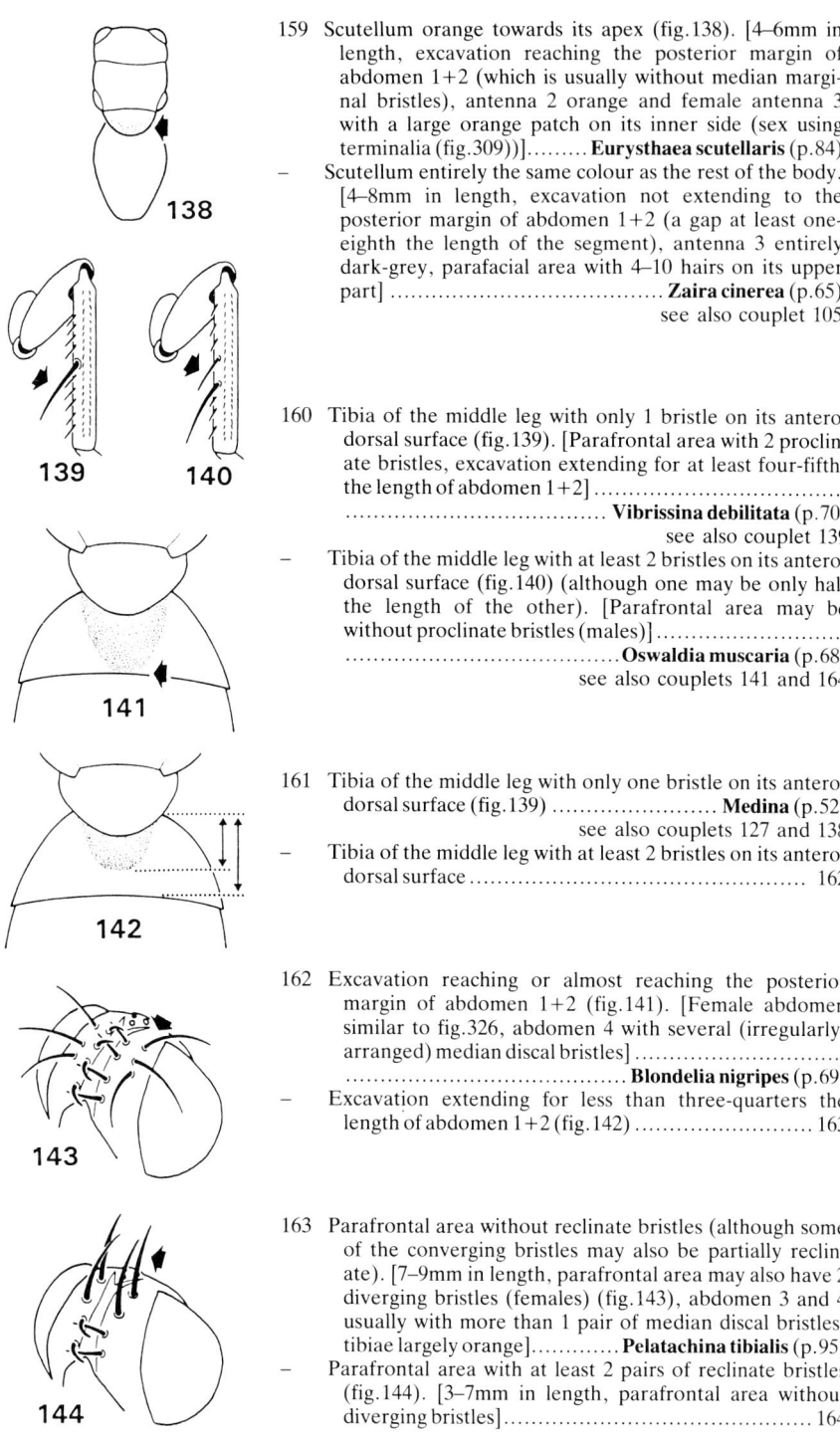

159 Scutellum orange towards its apex (fig.138). [4–6mm in length, excavation reaching the posterior margin of abdomen 1+2 (which is usually without median marginal bristles), antenna 2 orange and female antenna 3 with a large orange patch on its inner side (sex using terminalia (fig.309))]......... **Eurysthaea scutellaris** (p.84)

— Scutellum entirely the same colour as the rest of the body. [4–8mm in length, excavation not extending to the posterior margin of abdomen 1+2 (a gap at least one-eighth the length of the segment), antenna 3 entirely dark-grey, parafacial area with 4–10 hairs on its upper part] .... **Zaira cinerea** (p.65)
see also couplet 105

160 Tibia of the middle leg with only 1 bristle on its antero-dorsal surface (fig.139). [Parafrontal area with 2 proclinate bristles, excavation extending for at least four-fifths the length of abdomen 1+2] ....
.... **Vibrissina debilitata** (p.70)
see also couplet 139

— Tibia of the middle leg with at least 2 bristles on its antero-dorsal surface (fig.140) (although one may be only half the length of the other). [Parafrontal area may be without proclinate bristles (males)]...........................
.... **Oswaldia muscaria** (p.68)
see also couplets 141 and 164

161 Tibia of the middle leg with only one bristle on its antero-dorsal surface (fig.139) .... **Medina** (p.52)
see also couplets 127 and 138

— Tibia of the middle leg with at least 2 bristles on its antero-dorsal surface 162

162 Excavation reaching or almost reaching the posterior margin of abdomen 1+2 (fig.141). [Female abdomen similar to fig.326, abdomen 4 with several (irregularly-arranged) median discal bristles] ....
.... **Blondelia nigripes** (p.69)

— Excavation extending for less than three-quarters the length of abdomen 1+2 (fig.142) 163

163 Parafrontal area without reclinate bristles (although some of the converging bristles may also be partially reclinate). [7–9mm in length, parafrontal area may also have 2 diverging bristles (females) (fig.143), abdomen 3 and 4 usually with more than 1 pair of median discal bristles, tibiae largely orange]............. **Pelatachina tibialis** (p.95)

— Parafrontal area with at least 2 pairs of reclinate bristles (fig.144). [3–7mm in length, parafrontal area without diverging bristles].. 164

145

146

147

148

149

150

164 Thorax entirely black excluding grey markings on (and immediately posterior to) the postpronotal lobes (fig.145). [Postpronotal lobe with only 2 bristles, para-frontal area with at least 1 proclinate bristle (male and female), 3–5mm in length] .... **Leiophora innoxia** (p.67)

– Thorax with light-grey stripes (at least on its anterior part). [Postpronotal lobe with 3 bristles (arranged in a triangle), 5–7mm in length] ....
.... **Oswaldia muscaria** (p.68)
see also couplets 141 and 160

165 Abdomen 3 and 4 without median discal bristles 166
– Abdomen 3 and 4 with median discal bristles (fig.146).......
.... 169

166 Eye very large – more than 11 times the height of the gena (fig.147). [Thorax with 4 katepisternal bristles] ......
.... **Thecocarcelia acutangulata** (p.80)
– Eye not more than 7 times the height of the gena 167

167 Ocellar bristles absent. [Bend in the medial vein equal to a right angle, palp and scutellum orange towards their tips] .... **Drino lota** (p.77)
see also couplet 68
– A pair of ocellar bristles present (fig.148).................. 168

168 Tibia of the middle leg with only 1 bristle on its antero-dorsal surface (fig.149) (occasionally another one less than half its length nearer the femur) AND palp (fig.152) orange. [Male abdomen with orange markings on the sides, female antenna 3 with approximately half its area orange, 7–8mm in length] ....
.... **Townsendiellomyia nidicola** (p.80)
– EITHER tibia of the middle leg with at least 2 large bristles on its antero-dorsal surface OR palp dark-brown/black... 169

169 Scutellum with at least half its area orange (fig.138). [Female abdomen uniformly yellowish-grey when viewed with the naked eye (less distinct in males)]..... 170
– Scutellum without or with only a faint orange at its apex. [Abdomen black with grey markings] 171

170 Abdomen 3 with at least a pair of median discal bristles and usually without additional marginal bristles next to the median pair (fig.150). [Thorax with the katepister-nal bristles arranged as in fig.330, wing with basicosta dark-brown] .... **Erycilla ferruginea** (p.84)
– Abdomen 3 without or with very small median discal bristles (only half the thickness and length of the median marginal bristles) and often with 1 or 2 additional marginal bristles next to the median pair. [Thorax with the katepisternal bristles arranged as in fig.331, wing with basicosta orange] .... **Erycia furibunda** (p.80)
see also couplet 146

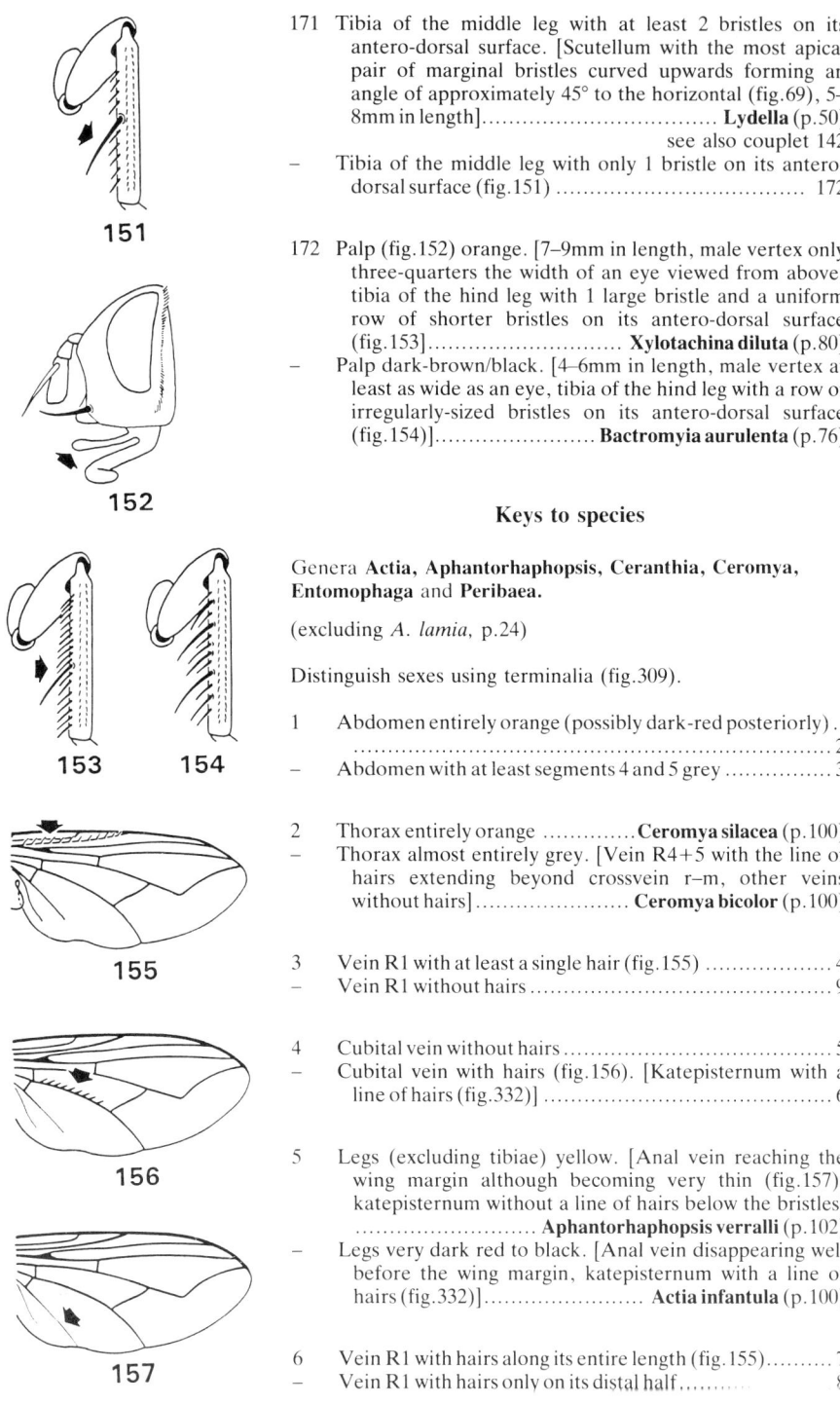

171 Tibia of the middle leg with at least 2 bristles on its antero-dorsal surface. [Scutellum with the most apical pair of marginal bristles curved upwards forming an angle of approximately 45° to the horizontal (fig.69), 5–8mm in length].................................. **Lydella** (p.50)
see also couplet 142

– Tibia of the middle leg with only 1 bristle on its antero-dorsal surface (fig.151) 172

172 Palp (fig.152) orange. [7–9mm in length, male vertex only three-quarters the width of an eye viewed from above, tibia of the hind leg with 1 large bristle and a uniform row of shorter bristles on its antero-dorsal surface (fig.153]............................ **Xylotachina diluta** (p.80)

– Palp dark-brown/black. [4–6mm in length, male vertex at least as wide as an eye, tibia of the hind leg with a row of irregularly-sized bristles on its antero-dorsal surface (fig.154)]....................... **Bactromyia aurulenta** (p.76)

Keys to species

Genera **Actia, Aphantorhaphopsis, Ceranthia, Ceromya, Entomophaga** and **Peribaea.**

(excluding *A. lamia*, p.24)

Distinguish sexes using terminalia (fig.309).

1 Abdomen entirely orange (possibly dark-red posteriorly)..
.... 2

– Abdomen with at least segments 4 and 5 grey 3

2 Thorax entirely orange ....**Ceromya silacea** (p.100)

– Thorax almost entirely grey. [Vein R4+5 with the line of hairs extending beyond crossvein r–m, other veins without hairs]....................... **Ceromya bicolor** (p.100)

3 Vein R1 with at least a single hair (fig.155) 4

– Vein R1 without hairs 9

4 Cubital vein without hairs 5

– Cubital vein with hairs (fig.156). [Katepisternum with a line of hairs (fig.332)] 6

5 Legs (excluding tibiae) yellow. [Anal vein reaching the wing margin although becoming very thin (fig.157), katepisternum without a line of hairs below the bristles] .... **Aphantorhaphopsis verralli** (p.102)

– Legs very dark red to black. [Anal vein disappearing well before the wing margin, katepisternum with a line of hairs (fig.332)]....................... **Actia infantula** (p.100)

6 Vein R1 with hairs along its entire length (fig.155).......... 7

– Vein R1 with hairs only on its distal half........... 8

151

152

153 **154**

155

156

157

7   Male antenna as in fig 334, female antenna 3 without or with only a small patch of orange at the junction with antenna 2 — extending for less than one-eighth its length (fig.158). [Ventral surface of vein R1 usually with hairs on its distal half, male terminalia as in fig.336]..............
.... **Actia crassicornis** (p.100)

–   Male antenna as in fig.335, female antenna 3 usually with a patch of orange extending for approximately half its length (fig.159). [Ventral surface of vein R1 usually without hairs, male terminalia as in fig.337].................
....**Actia pilipennis** (p.101)

158

8   Antenna 1, 2 and base of 3 (see fig.159) orange (rest of antenna 3 red) .... **Actia nudibasis** (p.101)

–   Antenna dark-grey (possibly a very narrow band of orange at the base of antenna 3) ....
.... **Actia maksymovi** (p.101)

159 pilipennis ♀

9   Vein R4+5 with the line of hairs extending approximately half-way between crossvein r–m and the wing margin (fig.160). [Male antenna 3 bilobed (fig.333) (unique among British Tachinidae)] ....
.... **Peribaea fissicornis** (p.101)

–   Vein R4+5 with hairs not extending beyond crossvein r–m (occasionally a single hair beyond) 10

160

10   Parafacial area with hairs on its entire length (fig.161) ....
..............................**Ceromya monstrosicornis** (p.100)

–   Parafacial area without hairs or with them confined to its upper half... 11

161

11   Anal vein reaching the wing margin although becoming very thin (fig.162)... 12

–   Anal vein disappearing well before the wing margin 13

162

12   Thorax with 3 post-sutural dorso-central bristles..............
.... **Ceranthia abdominalis** (p.102)

–   Thorax with 4 post-sutural dorso-central bristles (fig.163).
.... **Ceranthia lichtwardtiana** (p.102)

163

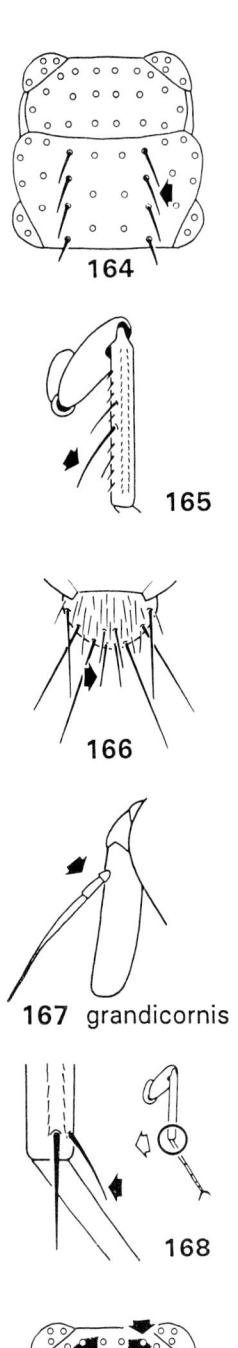

164

165

166

167 grandicornis

168

169

13   Thorax with 3 post-sutural dorso-central bristles. [Kate-pisternum with a line of hairs (fig.332)] ....
.... **Entomophaga exoleta** (p.99)

–   Thorax with 4 post-sutural dorso-central bristles (fig.164). [Katepisternum without a line of hairs below the bristles] .... **Entomophaga nigrohalterata** (p.99)

Genus **Admontia**

Distinguish sexes using terminalia (fig.309).
Characters distinguishing *grandicornis* and *maculisquama* (= *seria* of Emden (1954)) are not completely reliable.

1   Tibia of the middle leg with 2 or 3 large bristles on its antero-dorsal surface 2

–   Tibia of the middle leg with only 1 bristle on its antero-dorsal surface (fig.165) (occasionally another bristle nearer the femur but this is (usually much) less than two-thirds its length) 3

2   Scutellum with 3 pairs of large marginal bristles and a smaller pair at its apex which are however distinct from the surrounding hairs (fig.166), distal end of the tibia of the hind leg with the postero-dorsal bristle at least as long as the dorsal bristle, second segment of the arista approximately 3 times as long as broad (fig.167) ....
.... **grandicornis** (p.68)

–   Scutellum without apical bristles in addition to the 3 pairs of large marginal bristles, distal end of the tibia of the hind leg with the postero-dorsal bristle not more than two-thirds the length of the dorsal bristle (fig.168), second segment of the arista only approximately twice as long as broad........................ **maculisquama** (p.68)

3   Anterior part of the thorax with only a central light-grey stripe – the 2 rows of dorso-central bristles therefore on black (fig.169) (view from above and behind). [Parafacial area broad – the minimum distance between the eye and facial ridge 0.7 times (males) or equal to (females) the maximum width of antenna 3 at its mid-point] .... **blanda** (p.67)

–   Anterior part of the thorax with 3 parallel light-grey stripes. [Parafacial area only 0.4 (males) or 0.6 (females) times the width of antenna 3] .... **seria** (p.68)

Genus **Athrycia**

Distinguish sexes using terminalia (fig.309). External characters not completely reliable.

1   Proepisternum with hairs (fig.170), anatergite may have 2–4 short bristles/stout hairs (fig.171), arista starts to taper before its mid-point (fig.172). [Male sternite 5 with a pair of large lobes on its inner margin (fig.339)] .....
.... **curvinervis** (p.111)
–   Proepisternum without hairs (occasionally 1 or 2 present), anatergite without short bristles/stout hairs in position indicated, arista starts to taper only after its mid-point (fig.173).. 2

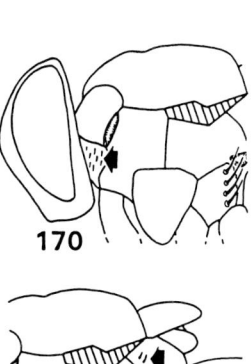

170

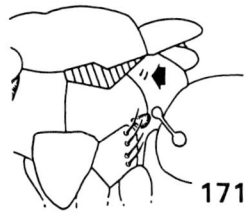

171

2   Wing with the distance between the bend in the medial vein and the wing margin 1.5 times the length of the medial vein after the bend (fig.342). [Male sternite 5 with a pair of small lobes on its inner margin (fig.340)]
.... **impressa** (p.112)
–   Wing with the distance between the bend in the medial vein and the wing margin not more than 1.3 times the length of the medial vein after the bend (fig.343) [Male sternite 5 without lobes on its inner margin (fig.341)] ......
.... **trepida** (p.112)

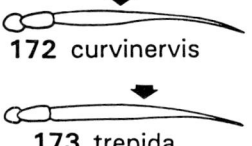

172 curvinervis

173 trepida

Genus **Bessa**

The single morphological character used to distinguish the species is not reliable.

1   Abdomen 3 and 4 with a pair of median discal bristles (fig.174)... **selecta** (p.64)
–   Abdomen 3 and 4 without median discal bristles ....
.... **parallela** (p.64)

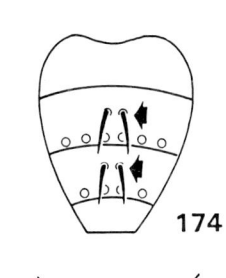

174

Genus **Blepharomyia**

Distinguish sexes using terminalia (fig.309). Characters not completely reliable.

1   Abdomen 1+2 with a complete row of marginal bristles – the ones adjacent to the median pair may be small but are distinct from the surrounding hairs (fig.175), vertex (fig.176) 1.2 times the width of an eye, male parafrontal area with some proclinate bristles of similar size to the convergent bristles .... **piliceps** (p.110)
–   Abdomen 1+2 without a complete row of marginal bristles, vertex 0.9 times (males) or equal to (females) the width of an eye, male parafrontal area with all proclinate bristles only approximately half the thickness and length of the convergent bristles ...... **pagana** (p.109)

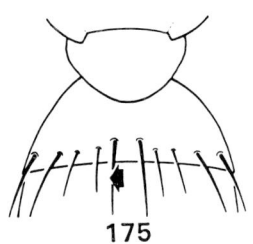

175

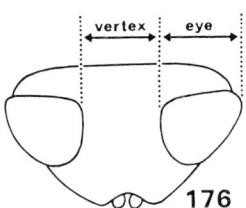

vertex   eye

176

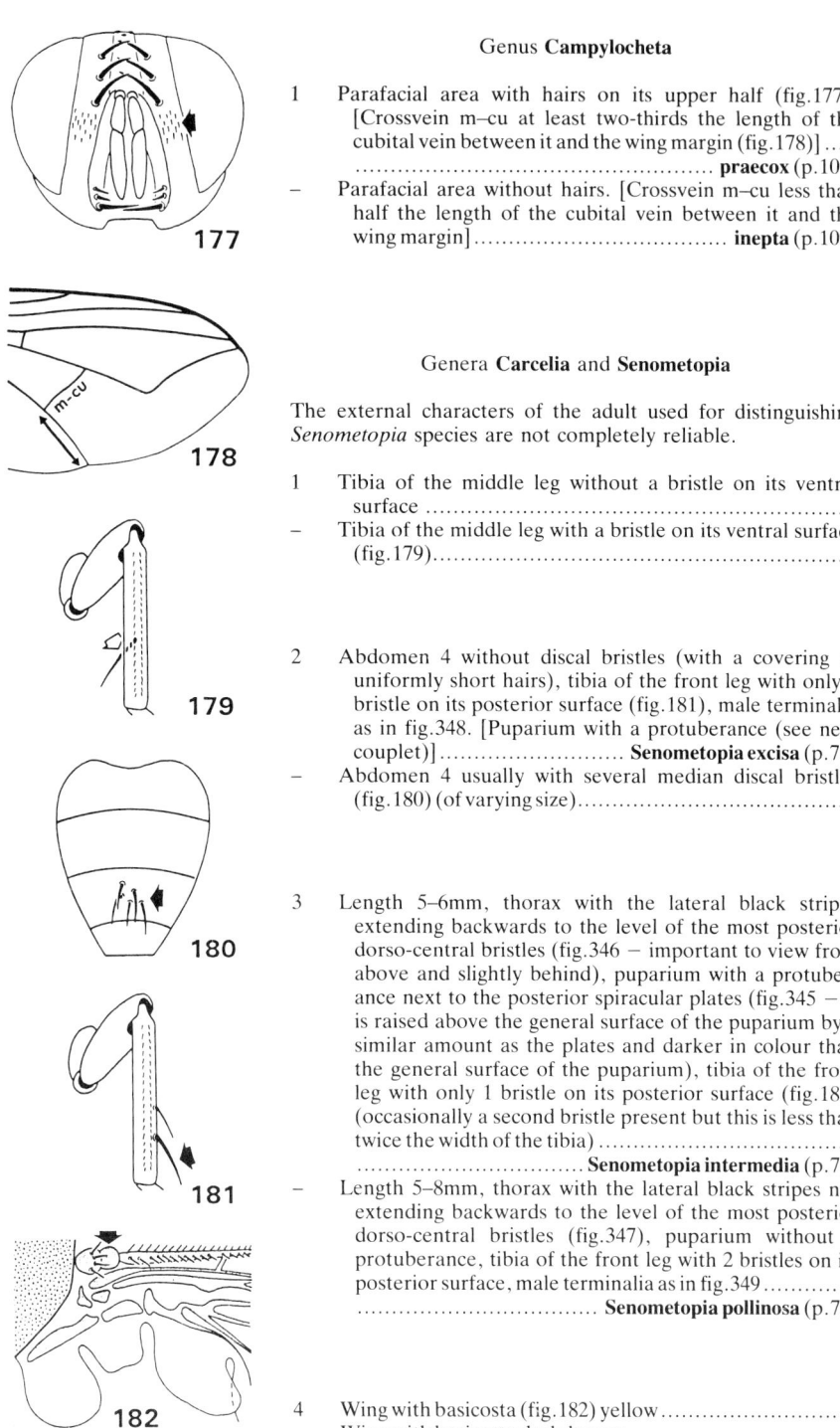

177

178

179

180

181

182

Genus **Campylocheta**

1　Parafacial area with hairs on its upper half (fig.177). [Crossvein m–cu at least two-thirds the length of the cubital vein between it and the wing margin (fig.178)] .... **praecox** (p.109)

–　Parafacial area without hairs. [Crossvein m–cu less than half the length of the cubital vein between it and the wing margin] .... **inepta** (p.109)

Genera **Carcelia** and **Senometopia**

The external characters of the adult used for distinguishing *Senometopia* species are not completely reliable.

1　Tibia of the middle leg without a bristle on its ventral surface 2

–　Tibia of the middle leg with a bristle on its ventral surface (fig.179).. 4

2　Abdomen 4 without discal bristles (with a covering of uniformly short hairs), tibia of the front leg with only 1 bristle on its posterior surface (fig.181), male terminalia as in fig.348. [Puparium with a protuberance (see next couplet)] .... **Senometopia excisa** (p.79)

–　Abdomen 4 usually with several median discal bristles (fig.180) (of varying size)...................................... 3

3　Length 5–6mm, thorax with the lateral black stripes extending backwards to the level of the most posterior dorso-central bristles (fig.346 – important to view from above and slightly behind), puparium with a protuberance next to the posterior spiracular plates (fig.345 – it is raised above the general surface of the puparium by a similar amount as the plates and darker in colour than the general surface of the puparium), tibia of the front leg with only 1 bristle on its posterior surface (fig.181) (occasionally a second bristle present but this is less than twice the width of the tibia) .... **Senometopia intermedia** (p.79)

–　Length 5–8mm, thorax with the lateral black stripes not extending backwards to the level of the most posterior dorso-central bristles (fig.347), puparium without a protuberance, tibia of the front leg with 2 bristles on its posterior surface, male terminalia as in fig.349 .... **Senometopia pollinosa** (p.79)

4　Wing with basicosta (fig.182) yellow 5

–　Wing with basicosta dark-brown 6

5   Vertex (fig.183) 0.42–0.50 (males) or 0.47–0.58 (females) times the width of an eye, abdomen 4 with hairs only two-fifths the length of the segment. [Male terminalia as in fig.350] .... **Carcelia rasa** (p.79)
–   Vertex 0.64–0.72 (males) or 0.72–0.80 (females) times the width of an eye, abdomen 4 with many hairs two-thirds the length of the segment. [Male terminalia as in fig.351] .... **Carcelia puberula** (p.79)

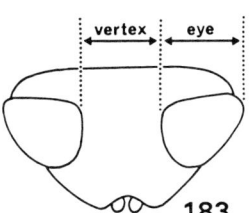
183

6   Abdomen 4 with several median discal bristles (fig.184) (of varying size) 7
–   Abdomen 4 without discal bristles 8

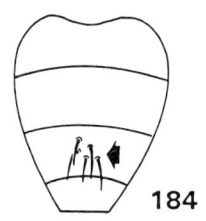
184

7   Tibia of the middle leg with only 1 bristle on its antero-dorsal surface (fig.185) (occasionally another nearer the femur but this less than half its length), scutellum usually with the most apical pair of marginal bristles not longer than the length of the scutellum, male terminalia as in fig.352 .... **Carcelia tibialis** (p.79)
–   Tibia of the middle leg usually with at least 2 large bristles on its antero-dorsal surface, scutellum usually with the most apical pair of marginal bristles 1.25 times the length of the scutellum, male terminalia as in fig.353 .... **Carcelia lucorum** (p.78)

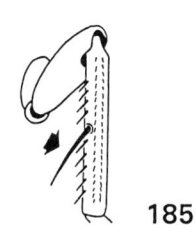
185

8   Arista starts to taper between two-fifths and half-way along its length (fig.186), tibia of the middle leg usually with 2 or more bristles on its antero-dorsal surface. [Male terminalia as in fig.355] .... **Carcelia gnava** (p.78)
–   Arista starts to taper one-third of the way along its length, tibia of the middle leg with only a single bristle on its antero-dorsal surface. [Male terminalia as in fig.354] .... **Carcelia atricosta** (p.78)

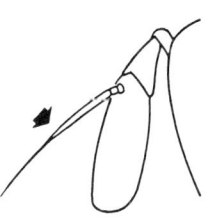

Genus **Cylindromyia**

1   Scutellum with 3 pairs of marginal bristles (fig.187). [9–11mm in length, abdomen without median discal bristles on segments 1+2, 3 and 4] ..... **brassicaria** (p.119)
–   Scutellum with 1 pair of marginal bristles. [6–8mm in length, abdomen with median discal bristles on segments 1+2, 3 and 4] .... **interrupta** (p.119)

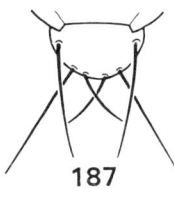
186 gnava

Genus **Dexia**

Distinguish sexes using presence (males) or absence (females) of yellow markings on abdomen.

1   Abdomen 4 with a black stripe along its posterior margin (fig.188) (females also with a similar stripe on abdomen 3). [6–9mm in length, thorax with 2 katepisternal bristles] .... **vacua** (p.108)
–   Abdomen without such markings. [7–12mm in length, thorax with 3 katepisternal bristles (occasionally 4)] .... **rustica** (p.108)

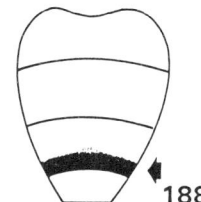
187

188

43

Genus **Dinera**

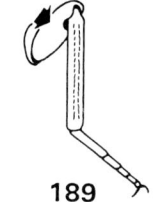

189

1   Femora (fig.189) may be mostly orange, abdomen uni-
     formly grey (view with the naked eye). [Abdomen 1+2
     without median marginal bristles, head as in fig.315] ......
     ....**grisescens** (p.107)
–   Femora entirely dark-brown/grey, abdomen black with
     light-grey markings. [Abdomen 1+2 may have a pair of
     median marginal bristles (males)] .... **carinifrons** (p.107)

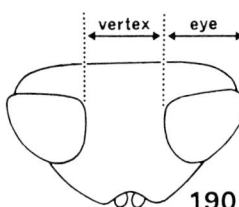

190

Genus **Dufouria**

Distinguish sexes using width of vertex (fig.190): equal to
three-quarters (females) or not more than one-quarter the
width of an eye (males).

No single character completely reliable (especially in females).

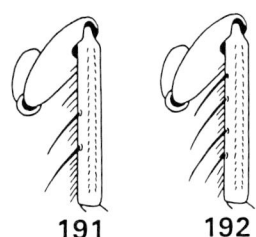

191   **192**

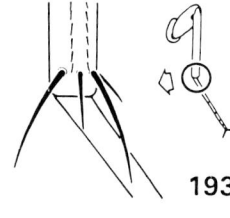

193

1   Tibia of the middle leg with only 2 bristles on its antero-
     dorsal surface (fig.191), tibia of the hind leg with only 2
     long bristles on its dorsal surface at the junction with the
     tarsus (fig.193) (the other 1 or 2 bristles are not more
     than half their length), male vertex narrower than the
     distance between the outer edges of the 2 posterior
     ocelli (fig.194). [4–5mm in length, male wing with
     neither of the 2 costal spines longer than crossvein r–m]...
     ....**nigrita** (p.114)
–   Tibia of the middle leg with at least 3 bristles on its antero-
     dorsal surface (fig.192), tibia of the hind leg with at
     least 3 long bristles on its dorsal surface at the junction
     with the tarsus, male vertex at least as wide as the
     distance between the outer edges of the 2 posterior
     ocelli. [5–6mm in length, male wing with costal spines
     often longer than crossvein r–m (these spines both in the
     position indicated in fig.313)] ....**chalybeata** (p.113)

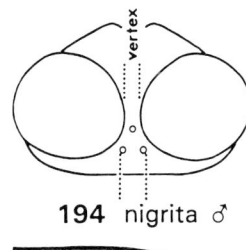

194 nigrita ♂

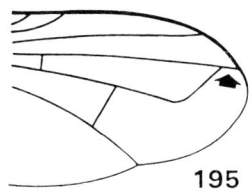

195

Genus **Eriothrix**

1   Wing with a petiole (fig.195). [Abdomen 3 with median
     discal bristles not longer than the maximum length of
     the segment, female abdomen with a pair of red spots on
     the sides (both wider than the gap separating them when
     viewed from above)] .... **rufomaculata** (p.109)
–   Wing without a petiole. [Abdomen 3 with median discal
     bristles 1.2–1.7 times the maximum length of the
     segment, female abdomen without red markings] ....
     .... **prolixa** (p.109)

Genera **Ernestia, Appendicia, Eurithia** and **Fausta**

The external characters used to distinguish *Eurithia* species are not completely reliable.

1 Parafacial area with (long, fine) hairs on its entire length (fig.196)..............................**Ernestia puparum** (p.91)
– Parafacial area without hairs or with them only above a point level with the mid-point of antenna 3.................2

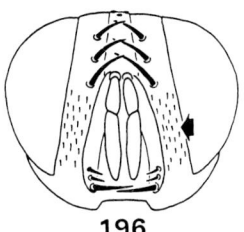

196

2 Scutellum with the most apical pair of marginal bristles arranged parallel or diverging – as long and thick as the adjacent pair...3
– Scutellum with the most apical pair of marginal bristles crossed – thinner and less than two-thirds the length of the adjacent pair (fig.197).....................................4

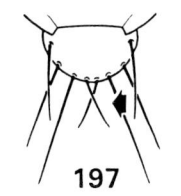

197

3 Medial vein with an appendix which is at least one-sixth the distance between crossvein m–cu and the bend (fig.198). [Scutellum without orange]..........................
.... **Fausta nemorum** (p.92)
– Medial vein without or with an appendix which is less than one-tenth the distance between crossvein m–cu and the bend. [Scutellum with at least the apical half orange]......
.... **Ernestia laevigata** (p.91)

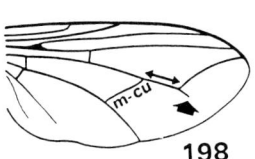

198

4 Female i.e. parafrontal area with 2 proclinate and 1 diverging bristle (fig.199).......................................5
– Male i.e. parafrontal area without proclinate or diverging bristles...6

199

5 Tarsus of front leg with third to last segment wider than long (see fig.356)..7
– Tarsus of front leg with third to last segment not wider than long (see fig.357)...8

6 Head without distinct inner vertical bristles (with an effort they may be found but they are not more than 1.25 times the length of the adjacent postocular hairs). [Vertex only one-fifth the width of an eye].................7
– Head with a pair of large inner vertical bristles (fig. 200) (at least twice as thick and long as the adjacent postocular hairs)...8

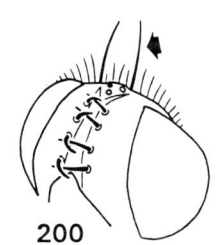

200

7 Thorax with the most posterior pair of (small) pre-sutural acrostichal bristles further from the suture than are the most posterior pre-sutural dorso-central bristles (fig.201). [The 2 rows of pre-sutural acrostichal bristles on a single central light-grey stripe which is usually as pronounced as the adjacent stripes the dorso-central bristles are on, male terminalia as in fig.361]................
....**Ernestia rudis** (p.91)
– Thorax with the most posterior pair of (small) pre-sutural acrostichal bristles nearer to the suture than are the most posterior dorso-central bristles. [Anterior part of thorax without or with only a faint central light-grey stripe, male terminalia as in fig.362]..........................
....**Ernestia vagans** (p.92)

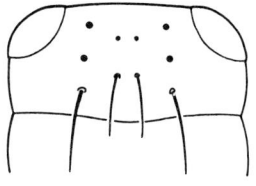

201 rudis

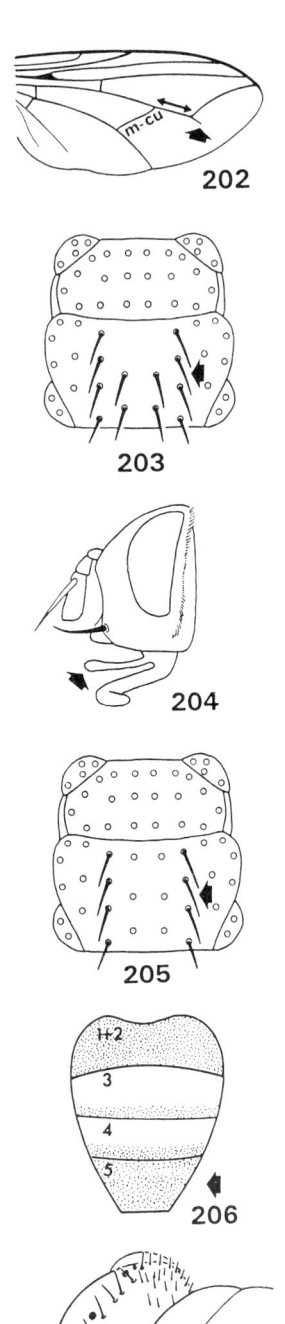

202

203

204

205

206

207

8   Medial vein with an appendix which is at least one-third the distance between crossvein m–cu and the bend (fig.202). [Thorax with 2 post-sutural acrostichal and 4 post-sutural dorso-central bristles (fig.203)].................
....**Appendicia truncata** (p.92)
–   Medial vein without or with an appendix which is not more than one-quarter the distance between crossvein m–cu and the bend. [Thorax usually with 3 post-sutural acrostichal bristles].. 9

9   Palp (fig.204) with more than the apical half orange...... 10
–   Palp dark-brown with at most the apical one-fifth lighter in colour 11

10   Thorax usually with 4 post-sutural dorso-central bristles (fig.205). [Male vertex 0.30–0.40 times the width of an eye viewed from above, male terminalia as in fig.363 (the large triangular process on the epandrium often visible without dissection)] ....
.... **Eurithia consobrina** (p.93)
–   Thorax with 3 post-sutural dorso-central bristles. [Male vertex 0.40–0.50 times the width of an eye, male terminalia as in fig.364 (note bifurcated process on epandrium)] ....**Eurithia vivida** (p.93)

11   Abdomen 5 without light-grey markings when viewed with the naked eye from above and behind (fig.206), head with most of the hairs behind the postocular row yellow (fig.207), [Thorax with 3 post-sutural dorso-central bristles, male vertex 0.55–0.65 times the width of an eye viewed from above, male terminalia as in fig.365]
......................................**Eurithia anthophila** (p.92)
–   Abdomen 5 with similar light-grey markings to those on abdomen 3 and 4, head with most of the hairs behind the postocular row white... 12

12   Antenna 2 at least as long as antenna 3 (fig.358). [Male terminalia as in fig.366, male vertex 0.45–0.60 times the width of an eye viewed from above, thorax usually with 4 post-sutural dorso-central bristles (fig.205)]..............
.... **Eurithia caesia** (p.93)
–   Antenna 2 shorter than antenna 3 13

13   Parafacial area broad (fig.359), male vertex at least three-quarters the width of an eye viewed from above. [Medial vein with a (short) appendix, thorax with 3 post-sutural dorso-central bristles, male terminalia as in fig.367].............................**Eurithia intermedia** (p.93)
–   Parafacial area narrow (fig.360), male vertex less than one-third the width of an eye. [Medial vein without an appendix, thorax with 4 post-sutural dorso-central bristles, male terminalia as in fig.368] ....
....**Eurithia connivens** (p.93)

Genus **Estheria**

1 Scutellum largely orange (fig.208). [Thorax with 3 post-sutural dorso-central bristles (fig.209) (occasionally only 2 or a smaller fourth present), wing with petiole not longer than crossvein r–m] .... **cristata** (p.108)
– Scutellum the same (grey) colour as the rest of the body. [Thorax with 4 post-sutural dorso-central bristles (fig.210), petiole may be longer than crossvein r–m] ....
.... **bohemani** (p.108)

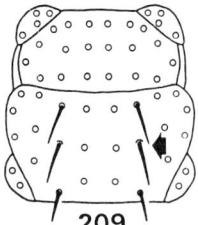

208

Genus **Exorista** (excluding *E.grandis* (p.24))

1 Scutellum with more than the apical one-quarter red (fig.208). [Thorax usually with 4 post-sutural dorso-central bristles (fig.210)] 2
– Scutellum without any red 3

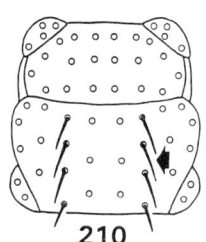

209

2 Abdomen 3 with the light-grey markings near the sides extending for at least three-quarters the length of the segment (fig.211 – ensure view at correct angle), male upper parafacial/lower parafrontal area with sparse hairs and with a gap between the 2 lines of bristles as indicated (fig.369). [Male terminalia with cercus narrowing abruptly near its tip (fig.371)]..... **larvarum** (p.61)
– Abdomen 3 with the light-grey markings near the sides extending for less than two-thirds the length of the segment, male upper parafacial/lower parafrontal area with denser hairs and without a gap between the 2 lines of bristles (fig.370). [Male terminalia with cercus narrowing gradually (fig.372)] .... **fasciata** (p.61)

210

211 larvarum

3 Thorax with 4 post-sutural dorso-central bristles (fig.210), abdomen 3 without marginal bristles adjacent to the median pair. [Male terminalia with a short ejaculatory apodeme (see fig.373)] .... **glossatorum** (p.62)
– Thorax with 3 post-sutural dorso-central bristles, abdomen 3 with a pair of marginal bristles adjacent to the median pair (fig.212). [Male terminalia with a longer ejaculatory apodeme (fig.374)] 4

4 Male terminalia (seen from below) with the cerci tapering to a point gradually (fig.375). [Abdomen 3 usually without a pair of median discal bristles] ... **tubulosa** (p.62)
– Male terminalia with the cerci coming to a point abruptly (fig.376). [Abdomen 3 usually with a pair of median discal bristles (fig.213)]...................................... 5

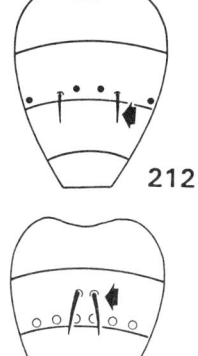

212

5 Male sternite 5 with a pair of lobes on its inner margin (fig.377)... **mimula** (p.62)
– Male sternite 5 without a pair of lobes on its inner margin (fig.378)... **rustica** (p.62)

213

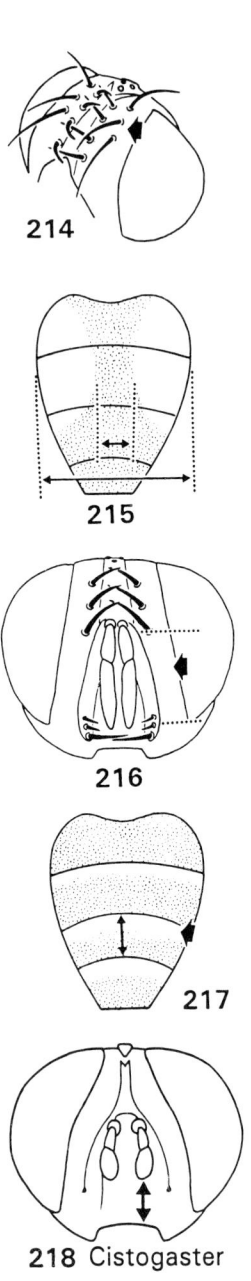

214

215

216

217

218 Cistogaster

219

Genus **Germaria**

1 Head as in fig.320. [9–12mm in length, parafrontal area with 2 proclinate and 1 diverging bristle (fig.214), postpronotal lobe entirely dark-brown/grey] ....
....**ruficeps** (p.89)
– Head as in fig.321. [6–8mm in length, parafrontal area may lack proclinate and diverging bristles (males), postpronotal lobe with approximately half its area yellow] .... **angustata** (p.88)

Genus **Gonia**

Occasional specimens not distinguishable.

1 Abdomen without yellow or red markings – occasionally faint red markings present which are separated by more than half the width of the abdomen (disappearing or becoming even fainter on the underside) ....
.... **picea** (p.87)
– Abdomen with large yellow or red markings – the intervening black stripe at its narrowest less than half the maximum width of the abdomen (fig.215).................. 2

2 Parafacial area (fig.216) appearing matt orange when viewed directly head-on with the naked eye (the angle is important), abdomen without or with narrow light-grey bands – extending for only one-eighth the length of abdomen 4.. **divisa** (p.86)
– Parafacial area appearing reflective yellowish-white, abdomen with broad light-grey bands – extending for more than one-quarter the length of abdomen 4 (fig.217).. 3

3 Abdomen with large yellow or red markings – the intervening black stripe at its narrowest less than one-fifth the maximum width of the abdomen, 9–13mm in length, tibiae brown.....................................**capitata** (p.86)
– Abdomen with smaller yellow or red markings – the intervening black stripe at its narrowest at least one-quarter the maximum width of the abdomen (fig.215), 8–10mm in length, tibiae very dark brown almost black
.... **ornata** (p.86)

Genera **Gymnosoma** and **Cistogaster**

1 Antenna only extending half-way to the oral cavity (fig.218). [EITHER anterior part of thorax with grey/gold markings which extend posteriorly beyond the thoracic suture (fig.220) (males) OR abdomen entirely black except possibly for red markings on segment 1+2 (occasionally extending slightly onto segment 3) (females), 4–5mm in length] ....**Cistogaster globosa** (p.116)
– Antenna extending more than three-quarters the way to the oral cavity. [Abdomen with yellow markings on all segments]............. 2

2   Length 6–8mm. [Anterior part of thorax with grey/gold markings which EITHER are restricted to on and around the postpronotal lobes (fig.219) (females) OR which extend posteriorly beyond the thoracic suture (fig.220) (males)] .... **Gymnosoma rotundatum** (p.116)

–   Length 3–4mm. [Anterior part of thorax EITHER entirely reflective black when viewed from above (females) OR with grey/gold markings which do not extend as far posteriorly as the thoracic suture (males)] .... **Gymnosoma nitens** (p.115)

220

Genus **Linnaemya**

Terminalia protruding from abdomen 5 in males but not in females.

221

1   Femora (fig.221) entirely orange (possibly small darker patches at their ends). [Abdomen largely orange – the central dark stripe not more than one-fifth the width of the abdomen]................................... **vulpina** (p.89)

–   Femora dark-brown/black. [Abdomen without (females) or with smaller orange markings (males) – the central dark stripe at least one-quarter the width of the abdomen].. 2

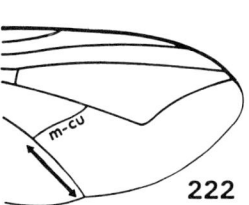

222

2   Crossvein m–cu only three-quarters the length of the cubital vein between it and the wing margin (fig.222). [Male parafrontal area with at least 1 proclinate bristle, male vertex at least as wide as an eye viewed from above]... **comta** (p.89)

–   Crossvein m–cu longer than the length of the vein between it and the wing margin. [Male parafrontal area without proclinate bristles, male vertex only two-thirds the width of an eye]... 3

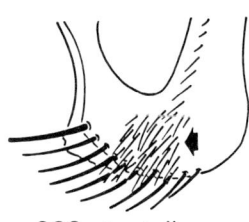

223 tessellans

3   Gena with only fine hairs (at least some of the lower ones also fair) – distinct from the single line of bristles extending down from the vibrissa (fig.223). [Costal cell of the wing (fig.225) with only approximately half its area covered in the microscopic hairs found over most of the wing] .... **tessellans** (p.90)

–   Gena with larger black hairs/small bristles – some of which are at least half the thickness of the bristles below the vibrissa (fig.224). [Costal cell with a uniform covering of microscopic hairs]................ **rossica** (p.90)

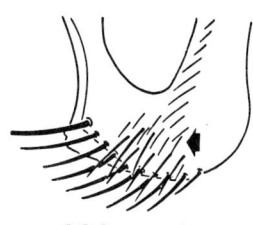

224 rossica

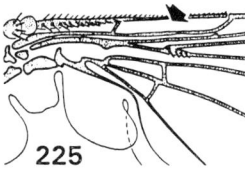

225

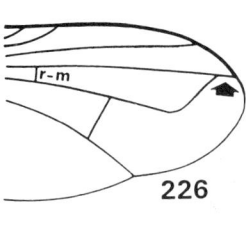

226

Genus **Loewia**

1   Wing with petiole at least twice the length of crossvein r–m. [Eye usually with long hairs, parafacial area with hairs at least on its upper part (usually extending down its entire length), 4–5mm in length] ... **submetallica** (p.95)

–   Wing with petiole approximately equal in length to crossvein r–m (fig.226). [Eye usually without or with hairs not longer than the maximum width of the arista] 2

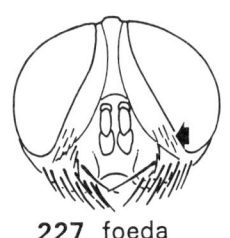

227 foeda

2   Parafacial area with hairs on its lower part (fig.227). [6–7mm in length, thorax usually with 4 post-sutural dorso-central bristles] .... **foeda** (p.95)

–   Parafacial area without hairs. [4–5mm in length, thorax with 3 post-sutural dorso-central bristles] ....
.... **phaeoptera** (p.95)

Genus **Lydella**

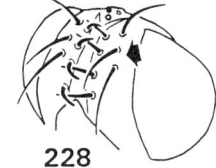

228

1   Male i.e. parafrontal area without proclinate bristles 2

–   Female i.e. parafrontal area with 2 proclinate bristles (fig.228) (may be difficult to distinguish further) 3

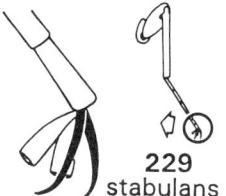

229
stabulans

2   Front and middle legs with tarsal claws approximately 1.25 times the length of the segment they are on (fig.229) (if broken the claws appear rectangular rather than pointed at their tips). [Arista starts to taper before or at its mid-point, vertex (fig.230) not wider than an eye] .... **stabulans** (p.77)

–   Tarsal claws only three-quarters the length of the segment. [Arista starts to taper usually only after two-thirds its length, vertex at least 1.3 times the width of an eye] .... **grisescens** (p.77)

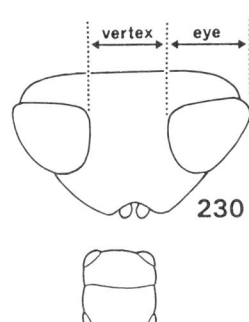

vertex   eye

230

3   Vertex (fig.230) not more than 1.2 times the width of an eye, arista starts to taper before or at its mid-point ....
.... **stabulans** (p.77)

–   Vertex at least 1.4 times the width of an eye, arista usually starts to taper only after two-thirds its length ....
.... **grisescens** (p.77)

Genus **Lypha**

♀ **231**

1   Scutellum with at least the apical third orange (fig.231). [Female abdomen 5 with an orange band along the posterior margin which extends for at least one-quarter the length of the segment (fig.231), wing membrane adjacent to crossvein r–m with a conspicuous dark-brown tinge] .... **ruficauda** (p.91)

–   Scutellum without orange. [Female without orange on abdomen 5, wing membrane adjacent to crossvein r–m not darker than that adjacent to the other veins] ....
..................,,,,,, ...**dubia** (p.90)

Genus **Macquartia**

1   Lower calypter running along the side of the scutellum (fig.232)...2
–   Lower calypter diverging away from the scutellum (fig.233). [5–8mm in length]....................................6

2   Legs entirely dark-brown. [Parafacial area with hairs confined to its upper part (fig.234), 5–8mm in length]..........
     .... **tenebricosa** (p.96)
–   At least the tibiae of the middle and hind legs orange. [Parafacial area with hairs extending down further, 6–11mm in length]....................................3

3   Female i.e. vertex more than three-quarters the width of an eye viewed from above. [Femora entirely orange]. (Some may not be further distinguishable).................4
–   Male i.e. vertex less than one-sixth the width of an eye ......
    5

4   Length of pre-alar bristle approximately 1.5 times the distance between its base and the postpronotal lobe (fig.235)... **viridana** (p.97)
–   Pre-alar bristle approximately equal to the distance between it and the postpronotal lobe .... **dispar** (p.96)

5   Femora entirely orange or with orange patches on their ventral surface extending for more than one-third their length. [Vertex usually at least one-sixth the width of an eye viewed from above] .... **viridana** (p.97)
–   Femora without orange. [Vertex usually not more than one-twelfth the width of an eye] .... **dispar** (p.96)

6   Halter (fig.236) brown. [Vein R4+5 with several hairs beyond the node, wing with basicosta dark-brown] ....
     .... **praefica** (p.96)
–   Halter light-yellow. [Vein R4+5 with hairs confined to the node (occasionally 1 beyond it)] 7

7   Parafacial area with hairs on its upper half (fig.234). [Abdomen may be uniformly grey (females) (view with the naked eye)].................................... **grisea** (p.96)
–   Parafacial area without hairs. [Abdomen black possibly with light-grey markings].....................................8

8   Node of vein R4+5 with at least 1 hair that is more than 1.5 times the length of crossvein r–m (fig.237). [Excavation usually only extending for half the length of abdomen 1+2].................................. **nudigena** (p.96)
–   Node of vein R4+5 only with hairs that are less than the length of crossvein r–m. [Palp and basicosta orange, excavation usually extending for at least three-quarters the length of abdomen 1+2] .... **pubiceps** (p.96)

232

233

234 tenebricosa

235 viridana

236

237

Genus **Medina**

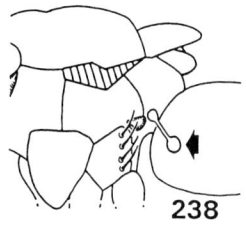

238

1 Halter (fig.238) with knob light-yellow. [4–8mm in length]
....**collaris** (p.66)
– Halter with knob dark-brown. [3–4mm in length]........... 2

2 Female i.e. head with inner vertical bristles at least twice
the thickness of the hairs in the postocular row (fig.239)
.... 3
– Male i.e. head with inner vertical bristles indistinct from
the hairs in the postocular row 4

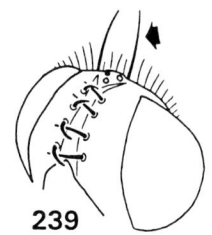

239

3 Postabdomen as in fig.385a .... **separata**
– Postabdomen as in fig.385b ....**luctuosa**

4 Sternite 5 with a dome-shaped arrangement of (mostly)
parallel hairs (fig.383 – dissection may be necessary)
....**luctuosa** (p.66)
– Sternite 5 with a spike-shaped arrangement of hairs drawn
together at their tips and a row of shorter hairs (fig.384)
....**separata** (p.66)

Genus **Meigenia** (excluding *M. majuscula*, p.20).

Only males distinguishable (by examination of terminalia).

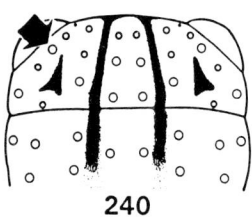

240

1 Surstylus with long hairs over its entire length (longer
than its width), cercus ends in a point (fig.433) ....
....**dorsalis** (p.65)
– Surstylus with only short hairs over most of its length,
cercus blunt and curved at its tip (fig.434) ....
....**mutabilis** (p.65)

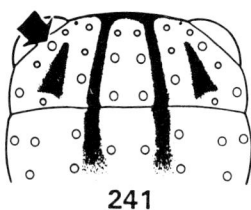

241

Genus **Myxexoristops**

1 Thorax with a pair of (well-defined) triangular black markings not extending forward beyond the bristle indicated
(fig.240) (view from above and behind), male scutellum
with at least a faint reddish tinge at its apex. [Male
terminalia with the tip of the cercus rectangular
(fig.435)]... **stolida** (p.82)
– Thorax with a pair of (less distinct) black markings extending forward beyond the bristle indicated (fig.241),
male scutellum without red. [Male terminalia with the
tip of the cercus rounded (fig.436)] ....**blondeli** (p.82)

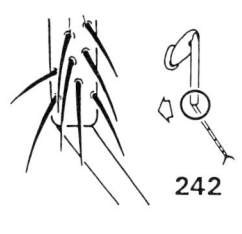

242

Genus **Phania**

1 Tibia of the hind leg with 8 to 10 bristles on its dorsal surface near the tarsus (fig.242). [5–7mm in length]............
....**thoracica** (p.119)
– Tibia of the hind leg with only 2 or 3 bristles on its dorsal
surface near the tarsus (fig.243). [3–5mm in length]........
................,,,,,..................................... **funesta** (p.119)

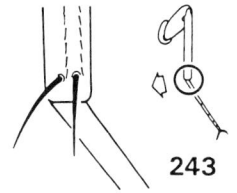

243

Genus **Phasia**

1   Thorax with the sides covered by long yellow hairs (view with the naked eye). [7–12mm in length, abdomen may have large orange markings] .... **hemiptera** (p.116)
–   Thorax without yellow hairs. [2–6mm in length, abdomen without yellow markings] 2

2   Thorax with light-grey markings when viewed from above and behind (fig.244). [4–6mm in length, petiole rarely more than twice the length of crossvein r–m] .... **obesa** (p.117)
–   Thorax entirely black. [2–5mm in length, petiole more than 3 times the length of crossvein r–m]...................... .... **pusilla** (p.117)

244

245

Genus **Phebellia**

The females of *vicina* and *villica* are not distinguishable (only 1 specimen (the Holotype) of *vicina* known (see Herting, 1981)).

1   Distance between the points where veins R2+3 and R4+5 reach the wing margin approximately the same as that between the latter and the wing tip (fig.392). [Distance between crossvein m–cu and the bend in the medial vein not greater than that between the bend and the nearest point on the wing margin (fig.245), male terminalia as in fig.387]... **glirina** (p.72)
–   Distance between the points where veins R2+3 and R4+5 reach the wing margin 1.5–2.0 times that between the latter and the wing tip (fig.393) 2

246

247

2   Facial ridge with hairs extending more than one-third of the way from the vibrissa to the lower margin of the base of the antenna (fig.246), parafacial area may have hairs (fig.247), abdomen 5 without light-grey markings or with them extending for less than half the length of the segment (fig.248) 3
–   Facial ridge with hairs not extending beyond one-third of the way from the vibrissa to the lower margin of the antennal base, parafacial area without or with only 1 or 2 hairs, abdomen 5 with light-grey markings which extend for more than half the length of the segment 4

248

3   Male terminalia as in fig.388 .... **vicina** (p.73)
–   Male terminalia as in fig.389 .... **villica** (p.73)

4   Female i.e. parafrontal area with 2 proclinate bristles (fig.249)... 5
–   Male i.e. parafrontal area without proclinate bristles 6

249

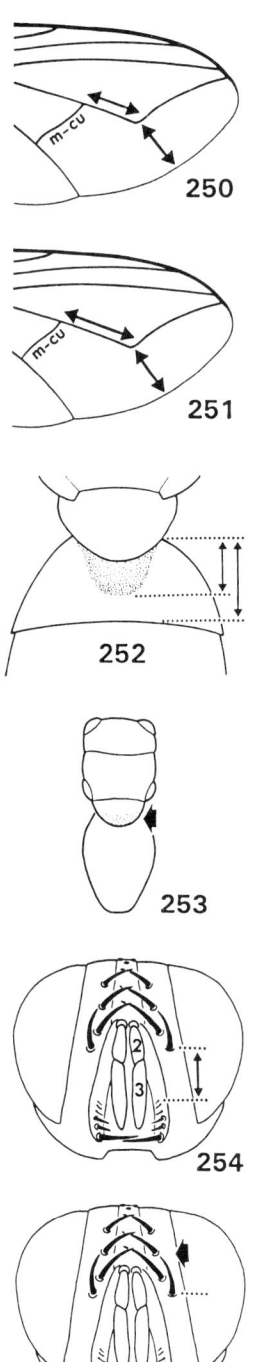

250

251

252

253

254

255

5   Distance between crossvein m–cu and the bend in the medial vein not greater than that between the bend and the nearest point on the wing margin (fig.250) ....
....**glauca** (p.72)

–   Distance between crossvein m–cu and the bend in the medial vein more than 1.5 times that between the bend and the nearest point on the wing margin ....
....**stulta** (p.73)

6   Distance between crossvein m–cu and the bend in the medial vein not greater than 1.5 times that between the bend and the nearest point on the wing margin (fig.251). [Terminalia as in fig.390] ....**glauca** (p.72)

–   Distance between crossvein m–cu and the bend in the medial vein approximately twice that between the bend and the nearest point on the wing margin. [Terminalia as in fig.391] ....**stulta** (p.73)

Genus **Phorocera**

1   Excavation almost reaching the posterior margin of abdomen 1+2 (the gap less than one-tenth the length of the segment), scutellum with at least the apical one-quarter (usually more) orange (fig.253). [7–13mm in length, male terminalia with the cerci as in fig.395 (view from below)] .... **assimilis** (p.63)

–   Excavation only extending for approximately two-thirds the length of abdomen 1+2 (fig.252), scutellum with little or no orange (only rarely extending over the apical one-quarter). [4–10mm in length, male terminalia with the cerci as in fig.396 (starting to taper closer to their apices)]..**obscura** (p.63)

Genus **Phryxe**

A few specimens of *nemea* and *vulgaris* not distinguishable.

1   A broad gap between the bristles extending up the facial ridge and those extending down from the parafrontal area – equal at least to two-thirds the length of antenna 3 (fig.254), parafrontal area (fig.255) yellowish-grey ......
....**nemea** (p.74)

–   Gap not more than half the length of antenna 3, parafrontal area bluish-grey.......................................2

2   Wing with at least 1 hair on the underside of the subcostal vein (fig.394). [Male terminalia as in fig.397] ....
....**heraclei** (p.74)

–   Wing without hairs on the underside of the subcostal vein ..
....3

3   Male terminalia as in fig.398. [Scutellum usually with only its apical one-quarter or less orange] ....
....**magnicornis** (p.74)

–   Male terminalia as in fig.399. [Scutellum usually more extensively orange (fig.253)] ....**vulgaris** (p.75)

Genus **Siphona** (males only)

Distinguish sexes using terminalia (fig.309). Partial identification of females possible using characters excluding abdominal colour, tarsal claw length and antennal length.
For measurements see fig.403.

1   Thorax with 3 post-sutural dorso-central bristles............ 2
–   Thorax with 4 post-sutural dorso-central bristles (fig.256).
      [Haustellum 1.0–1.2 times the height of the head] 7

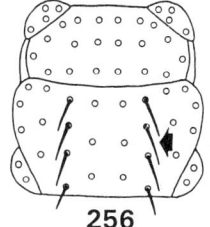

256

2   Tarsal claws as long as the tarsal segment to which they
      are attached (fig.257 – unique among British *Siphona*).
      [Abdomen without orange markings (unique except for
      some *geniculata*), distiphallus with very large apical
      teeth (fig.413), haustellum 0.9 times the height of the
      head, head profile as in fig.406] .... **ingerae** (p.103)
–   Tarsal claws only half the length of the segment 3

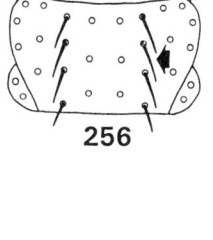

257

3   Abdomen 1+2 without marginal bristles 4
–   Abdomen 1+2 with a pair of marginal bristles at the sides
      of the segment (fig.258) 5

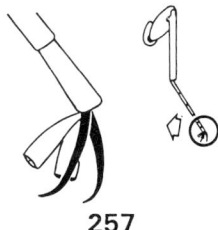

258

4   Thorax with a pair of pre-sutural intra-alar bristles
      (fig.259). [Haustellum only 0.8 times the height of the
      head (unique among British *Siphona*), distiphallus with
      large apical teeth (fig.414), antenna 1.4–1.8 times the
      length of the frons (fig.407)] ....**maculata** (p.103)
–   Thorax without pre-sutural intra-alar bristles (unique
      among British *Siphona*). [Haustellum 0.85–0.95 times
      the height of the head, distiphallus with smaller apical
      teeth (fig.415), antenna 1.2–1.5 times the length of the
      frons (head profile similar to *mesnili*)].......**collini** (p.102)

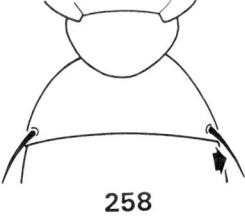

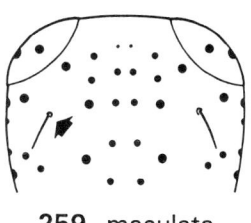

259 maculata

5   Femur of the front leg with 4 bristles on its postero-
      ventral surface (fig.404 – unique among British *Siphona*
      except for *ingerae*). [Sternite 5 with the apex of the lobes
      turned abruptly inwards forming a 90° angle (fig.401 –
      unique), distiphallus as in fig.416, antenna 1.4–1.7 times
      the length of the frons (fig.408)] .... **mesnili** (p.103)
–   Femur of the front leg with a maximum of 3 bristles on its
      postero-ventral surface (fig.405). [Sternite 5 with the
      apex of the lobes forming a much greater angle
      (fig.402)].. 6

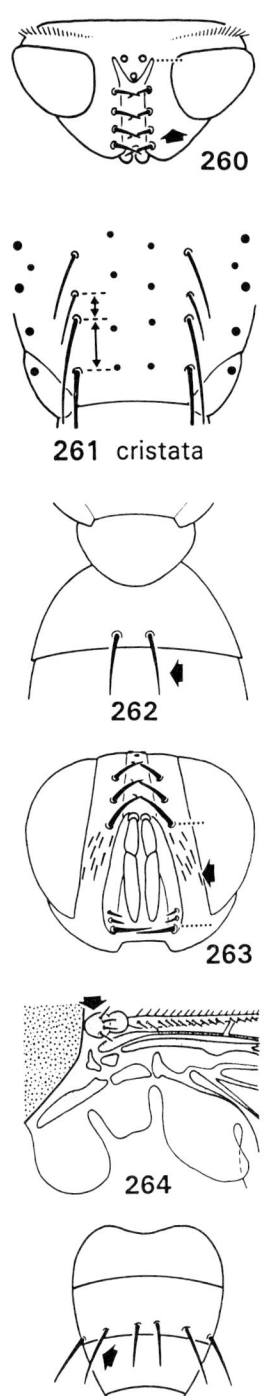

260

261 cristata

262

263

264

265

6 Parafrontal area (fig.260) yellower than the thorax (unique among British *Siphona* – view from above under low magnification), palp without hairs on its distal one-sixth. [Head profile similar to *geniculata*, average width of head seen from above = 1.43mm] ....
....**cristata** (p.103)
see also couplet 8
– Parafrontal area not yellower than the thorax, [?Small species (average width of head in 6 specimens examined = 1.14mm), antenna 1.3–1.5 times the length of the frons (head profile similar to *mesnili*), distiphallus as in fig.418] .... **variata** (p.104)

7 Antenna only 1.0–1.3 times the length of the frons (fig.403)...8
– Antenna more than 1.3 times the length of the frons. [Parafacial area with a maximum of 6 hairs (which are not longer than the maximum width of the arista)] 9

8 Distance between the two most posterior post-sutural dorso-central bristles twice that between the middle two (fig.261). [Distiphallus broad (fig.417)] ....
....**cristata** (p.103)
see also couplet 6
– Distance between the two most posterior post-sutural dorso-central bristles not more than 1.5 times that between the middle two. [Abdomen 1+2 may have a pair of median marginal bristles (fig.262 – unique among British *Siphona* except for *setosa*), parafacial area may have hairs extending down to the mid-point between the lowest parafrontal bristles and the vibrissa (fig.263) (unique), distiphallus slender (fig.419), head as in fig.409] .... **geniculata** (p.103)

9 Abdomen 1+2 with a pair of median marginal bristles (fig.262) (unique among British *Siphona* except for some *geniculata*). [Abdomen 3 with 3 pairs of marginal bristles (fig.265), tegula (fig.264) light-red, antenna 1.5–1.7 times the length of the frons (fig.410), distiphallus as in fig.420] ....**setosa** (p.104)
– Abdomen 1+2 without median marginal bristles. [Abdomen 3 with only 2 pairs of marginal bristles] 10

10 Antenna 1.3–1.5 times the length of the frons (fig.411), distiphallus with teeth in the area arrowed in fig.421. [Small species – average width of head seen from above = 1.26mm] .... **pauciseta** (p.103)
– Antenna 1.5–1.7 times the length of the frons (head profile similar to *setosa* except antenna 3 broader), distiphallus as in fig.422. [Tegula (fig.264) dark-brown/black, average width of head = 1.38mm] ....
....**boreata** (p.102)

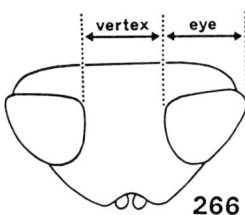

Genus **Solieria** (males only)

Distinguish sexes using terminalia (fig.309). Females and occasional males not distinguishable.

1   Vertex (fig.266) at least 0.9 times the width of an eye, parafrontal area usually with proclinate bristles at least on 1 side (fig.267) 2
–   Vertex not more than 0.7 times the width of an eye, parafrontal area without proclinate bristles...................... 3

2   Tarsal claws only approximately two-thirds the length of the tarsal segment to which they are attached, femur of the front leg with at least the basal one-quarter of its ventral surface dark red/brown (fig.269), parafrontal area with at least 2 proclinate bristles on both sides (fig.267)....................................... **pacifica** (p.105)
–   Tarsal claws as long as the segment (fig.268), femur of the front leg with its ventral surface entirely yellow, parafrontal area may be without or with only a single proclinate bristle .... **vacua** (p.105)

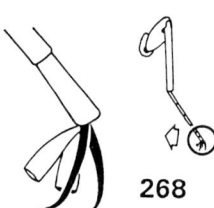

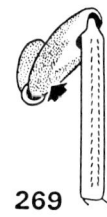

3   Vertex (fig.266) more than 0.6 times the width of an eye, femur of the front leg with at least the basal one-quarter of its ventral surface dark red/brown (fig.269) ....
....**fenestrata** (p.105)
–   Vertex less than 0.5 times the width of an eye, femur of the front leg with its ventral surface entirely yellow ....
....**inanis** (p.105)

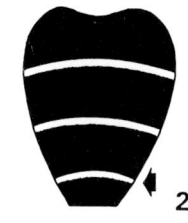

Genus **Tachina** subgenus **Servillia**

1   Abdomen 5 with a white band across its anterior margin (fig.270), abdomen usually without orange markings (view from above with the naked eye). [Parafacial area usually with most of the hairs black (fig.271), male antenna 3 only approximately 1.25 times as broad as antenna 2] ....**ursina** (p.88)
–   Abdomen 5 without a white band across its anterior margin (possibly present on the sides of the segment only), abdomen with orange markings. [Parafacial area usually with most of the hairs white, male antenna 3 at least 1.5 times as broad as antenna 2 (fig.271)] ....
....**lurida** (p.88)

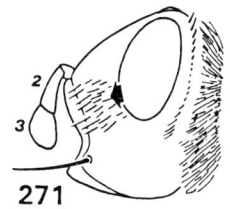

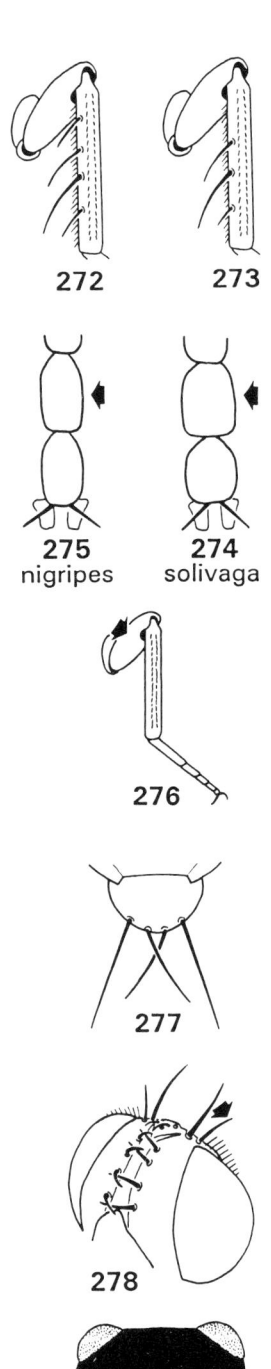

272   273

275   274
nigripes solivaga

276

277

278

279

Genus **Thelaira**

Occasional specimens not distinguishable.

1   Tibia of the middle leg with 4 long bristles on its antero-dorsal surface (fig.272), male abdomen with large orange markings — the intervening black stripe only approximately one-third the maximum width of the abdomen, female tarsus with the second to last segment only 1.2–1.4 times as long as broad (fig.274) .... **solivaga** (p.113)

–   Tibia of the middle leg with only 2 long bristles on its antero-dorsal surface (fig.273) (possibly also a shorter bristle present nearer the tarsus), male abdomen with smaller orange markings — the intervening black stripe approximately half the maximum width of the abdomen, female tarsus with the second to last segment 1.5–2.0 times as long as broad (fig.275) .... **nigripes** (p.113)

Genus **Trixa**

1   Femora (fig.276) entirely orange (occasionally darker on their dorsal surface). [Crossvein r–m orange, scutellum with the most apical pair of marginal bristles extending parallel to each other].................. **caerulescens** (p.107)

–   Femora almost entirely dark red-brown or grey. [Crossvein r–m and the adjacent wing membrane dark-brown, scutellum usually with the most apical pair of marginal bristles crossed] .... **conspersa** (p.106)

Genera **Wagneria, Periscepsia** and **Ramonda** (excluding *R. prunaria*, p.25)

1   Scutellum with only 2 pairs of large marginal bristles (fig.277) (occasionally some others which are less than half the length of the apical pair). [Outer vertical bristles twice the length of the adjacent postocular hairs (fig.278)]..................... **Periscepsia carbonaria** (p.111)

–   Scutellum with 3 pairs of large marginal bristles 2

2   Anterior part of the thorax with light-grey markings (view from above and behind). [Outer vertical bristles not distinguishable from the adjacent hairs in the postocular row — therefore only 1 pair of bristles (the inner vertical pair) behind the ocelli]... 3

–   Thorax entirely reflective black excluding the post-pronotal lobes which may be grey (fig.279). [Outer vertical bristles at least half the length of the inner vertical pair (fig.278)]... 4

3　Head with a second (smaller) pair of ocellar bristles (fig.280). [Wing with petiole rarely more than 1.5 times the length of crossvein r–m, basicosta brown] ....
.... **Ramonda spathulata** (p.110)

–　Head with only 1 pair of ocellar bristles. [Petiole at least twice the length of crossvein r–m, basicosta yellow] ....
.... **Ramonda latifrons** (p.110)

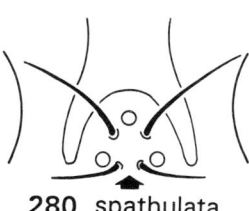

280 spathulata

4　Abdomen entirely reflective black. [Palp (fig.281) orange/light-brown, parafacial area with at the most 1 or 2 hairs in addition to the bristles] .... **Wagneria costata** (p.111)

–　Abdomen with light-grey markings (view from above and behind). [Palp dark-brown/black, parafacial area with hairs in addition to the bristles] ....
.... **Wagneria gagatea** (p.111)

281

Genus **Winthemia**

1　Upper part of the head also with black hairs behind the postocular row (fig.282). [Abdomen 3 and 4 may have (irregular) median discal bristles, abdomen 5 may lack orange markings, male terminalia as in fig.438] ....
.... **variegata** (p.71)

Upper part of the head with only white hairs behind the postocular row. [Abdomen 3 and 4 without median discal bristles (with a covering of uniformly short hairs), abdomen 5 with a broad orange band along its posterior margin (fig.283)] 2

282

283

2　Anterior part of the thorax with a pair of broad black stripes – each at least one-third (males) or one-sixth (females) the width of the intervening space (fig.284), male abdomen 4 with the marginal bristles longer than the segment, female hind tibia with the longest bristle on its postero-dorsal surface 2.0–3.5 times the width of the tibia (fig.285). [Male terminalia as in fig.439 (cercus with a short tooth-like apex)]...................................
.... **quadripustulata** (p.71)

–　Anterior part of the thorax with a pair of narrower black stripes – each not more than one-quarter (males) or one-sixth (females) the width of the intervening space, male abdomen 4 with the marginal bristles shorter than the segment, female hind tibia with the longest bristle on its postero-dorsal surface only 1.0–1.8 times the width of the tibia. [Male terminalia as in fig.440 (surstylus broad)]............................**cruentata** (p.71)

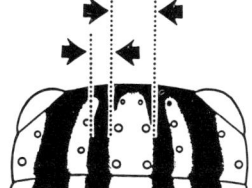

284 quadripustulata
♂

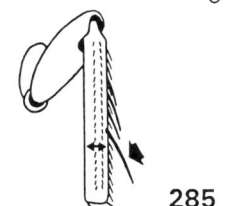

285

Species biology

The hosts, distribution, habitat and flight period of each species are summarised here (as far as they are known). Other biological data are largely from Herting (1960).

During preparation of this work the collections at the following institutions were examined:

NHM – Natural History Museum, London (including Fonseca and Spooner collections).

UMO – University Museum, Oxford (including Collin/Verrall and Varley/Oak Survey collections).

BCM – Bristol City Museum and Art Gallery (including the Audcent collection). A few records from Cambridge University Zoology Department may be accidentally included amongst these.

UMM – University Museum, Manchester.

LM – Liverpool Museum.

NMW – National Museum of Wales, Cardiff.

Additional data (mostly Irish) have been included based on material examined by Mr P.Chandler, mostly at the following institutions.

NMD – National Museum, Dublin.

SI – Smithsonian Institution (Stelfox collection), Washington.

BENHS – British Entomological and Natural History Society collection.

Royal Scottish Museum, Edinburgh.

Hosts. All host records found by the author are given with their source or, in the case of multiple records, with their number. Where only a number is given, at least one of the tachinid specimens has been seen by the author. Rearings from the same host, at the same locality, in the same year have been treated as single records. Some records in Audcent (1942) have not been treated as British as they lack collection data.

Host records outside Britain are given in summarised form only. These are taken from the host-list of Palaearctic Tachinidae currently being compiled by Dr H-P. Tschorsnig and Dr B. Herting at the Staatliches Museum für Naturkunde, Stuttgart. Where there are only a few records, the number is also given.

Nomenclature of British hosts follows Bradley & Fletcher (1986) (Lepidoptera) and the relevant R.E.S.L. checklists (other taxa).

British distribution. The distribution of species within the British Isles is indicated using the regions shown in fig.304. Counties are given only if the species has been recorded from fewer than a minimum number of counties within the region. These minimum numbers are: 6 for S.England; 3 for Wales, the Midlands and N. Scotland; 2 for N.England. All S.Scotland and Irish counties are given. Patterns within these regions are indicated. An asterisk denotes a literature record.

The distributions given are of recorded presence only and not probable presence or absence. The collections examined are biased towards S.England and a few localities in N.Scotland (the absence from S.Scotland of many species recorded from N.Scotland is probably due to undercollecting).

Habitat. Herting (1960) gives habitat data for many species in continental Europe (central and northern) and Falk (in press) has examined the collection localities of all the rarer British species. The habitat data in Emden (1954) (taken largely from Day (1948)) are questionable. They are not repeated here unless they indicate a restricted habitat preference.

Flight period. For each species a calendar period is given which includes 90% of the records in the collections examined by the author. The term "early" refers to the 1st to the 15th of the month; "late" refers to the 16th to the end of the month. The number of records upon which this is based is given in brackets. Peaks within this period are also indicated.

Subfamily Exoristinae

This subfamily contains a number of different reproductive strategies (see under tribes).

Tribe Exoristini

Species in this tribe lay unincubated eggs on Lepidoptera, Coleoptera and sawfly (Hymenoptera: Symphyta) larvae.

Genus **Exorista**

E. fasciata. May be gregarious.
Hosts: Lepidoptera larvae, mainly Lymantriidae and Lasiocampidae. In Britain: Arctiidae − *Arctia caja* (3), *Callimorpha dominula* (NMW), *Tyria jacobaeae* (UMM). Lasiocampidae − *Lasiocampa trifolii* (3), *L.quercus* (UMO), *Macrothylacia rubi* (14), *Malacosoma castrensis* (UMO), *Philudoria potatoria* (3). Lymantriidae − *Dicallomera fascelina* (Ford & Shaw, 1991), *Euproctis similis* (UMO). Zygaenidae − *Zygaena filipendulae* (9), *Z.trifolii* (NHM).
British distribution: S.England (south of a Bristol–London line), Midlands (Cheshire), Wales, N.England, S.Scotland (Selkirk and Ayr*), N.Scotland (Sutherland and Inverness*) and Ireland (Cork, Donegal, Galway, Kerry, Dublin, Clare and Sligo).
Habitat: in Europe usually found in extreme biotopes − coastal sandhills, dry heath and pine woods (Herting, 1960).
Flight period: late May to late August (at least 50 records). Probably 2 generations per year.

E. larvarum. Usually gregarious, overwintering in the host and pupating in the host cocoon or the soil.
Hosts: Lepidoptera larvae, mainly Lymantriidae and Lasiocampidae. In Britain: Arctiidae − *Arctia villica* (UMO), *Parasemia plantaginis* (Ford & Shaw, 1991), *Phragmatobia fuliginosa* (NHM), *Tyria jacobaeae* (Ford & Shaw, 1991). Lasiocampidae − *Philudoria potatoria* (6), *Trichiura crataegi* (Ford & Shaw, 1991). Lymantriidae − *Dicallomera fascelina* (UMO). Noctuidae − *Acronicta rumicis* (NHM).
British distribution: S.England, Midlands (Worcs), Wales, N.England, S.Scotland (Ayr*), N.Scotland (Argyll and Sutherland*) and Ireland (Kerry, Galway, Wicklow and Waterford).

Flight period: late July to early September (80% of records) but some May and June (at least 50 records). Similar pattern in Germany (Tschorsnig, 1989). Probably 2 generations per year.

E. glossatorum

Hosts: unknown.
British distribution: S.England (Devon, Dorset and Hants).
Habitat: localities indicate broadleaved woodland (Falk, in press).
Flight period: late June/early July (11 records).

E. grandis

Hosts: mainly larvae of *Pavonia pavonia*, but also other Lepidoptera. In Britain: Arctiidae − *Arctia caja* (NHM). Lymantriidae − *Dicallomera fascelina* (UNO). Pieridae − *Pieris brassicae* (NHM). Saturniidae − *Pavonia pavonia* (34).
British distribution: S.England, Midlands, Wales (Cardigan and Glamorgan), N.England, S.Scotland (Roxburgh*), N.Scotland and Ireland (Cork).
Habitat: usually heathland (Herting, 1960).
Flight period: late June to late August (18 records). In Europe 2 generations per year (Herting, 1960).

E. mimula

Hosts: in Europe sawfly larvae (Hymenoptera: Symphyta): Tenthredinidae − *Cladius pectinicornis* (2), *Athalia rosae* (4) and *Pristiphora pallidiventris* (1).
British distribution: S.England (Cornwall, Devon, Dorset, Cambridge and Kent) and Wales (Glamorgan). There are additional English and Welsh records of either *rustica* or *mimula*.
Flight period: late July and August (17 records).

E. rustica. Overwinters as a first-instar larva within the host.

Hosts: in Europe Tenthredinidae sawfly larvae (Hymenoptera: Symphyta). In Britain from *Dolerus* sp. (Tenthredinidae) (NHM).
British distribution: S.England, Midlands, Wales (Monmouth), N.England, S.Scotland (Dumfries), N.Scotland (Perth) and Ireland (Kerry*, Galway, Dublin, Mayo and Antrim*).
Flight period: late June to late September (at least 50 records). More than 1 generation per year.

E. tubulosa

Hosts: unknown.
British distribution: S.England, Midlands (Hereford) and Wales (Radnor).
Habitat: downland (Emden, 1954).
Flight period: early June to late August (20 records).

Genus **Chetogena**

C. acuminata

Hosts: in the Soviet Union and Yugoslavia larvae of several species of Tenebrionidae (Coleoptera).
British distribution: S.England (Winterton sandhills and Yarmouth Denes in Norfolk; Felixstowe in Suffolk).
Habitat: localities indicate coastal dunes (Falk, in press). A southern European

and North African species which is also found on sand dunes in southern England and Holland (Herting, 1960).

Flight period: late July and early August (5 records).

Genus **Diplostichus**

D. janithrix. May be gregarious, pupating in the host cocoon.

Hosts: sawfly larvae in the genus *Diprion* (Hymenoptera: Symphyta: Diprionidae), the majority of records from *D.pini*. In Britain 5 records from this species and 1 from an unidentified sawfly on *Epilobium* (UMO). In the Soviet Union there is a questionable record from *Neodiprion sertifer*.

British distribution: S.England (Dorset, Hants, Surrey, Suffolk and Norfolk).

Habitat: localities indicate woodland and heathland (Falk, in press). The host genus feeds on pine.

Flight period: late July to late September (9 records). Similar in Europe (Herting, 1960).

Genus **Parasetigena**

P. silvestris. Usually solitary, the mature maggot emerging from the host prepupa or pupa and overwintering as a puparium in the soil.

Hosts: in Europe *Lymantria dispar* and *L.monacha* larvae (Lepidoptera: Lymantriidae). A single British record from *Polyploca ridens* (Lep.: Thyatiridae) (UMO) is probably erroneous.

British distribution: S.England (Lyndhurst and Rhinefield both in the New Forest in Hants; Ham Street in Kent) and Midlands (unknown locality in Hcreford*).

Habitat: localities indicate old broadleaved woodland (Falk, in press).

Flight period: late May and early June (5 records). Similar pattern in Germany (Tschorsnig, 1989). 1 generation per year.

Genus **Phorocera**

The eggs and female terminalia are specially modified for partial insertion of the egg into the host integument (Herting, 1963; Wood, 1972).

P. assimilis

Hosts: in Europe mostly tree-feeding Lepidoptera larvae, mainly Geometridae and Noctuidae. In Britain: *Cosmia trapezina* and *Orthosia cruda* (both Noctuidae) on *Quercus* (UMO).

British distribution: S.England and Midlands.

Habitat: deciduous woodland (Herting, 1960).

Flight period: May and early June (at least 50 records).

P. obscura. Solitary, leaving the host to pupate in the soil.

Hosts: tree-feeding Geometridae larvae (Lepidoptera), and a few tree-feeding Noctuidae (Lepidoptera). In Britain: Geometridae – *Agriopis leucophaearia* (NHM), *Apocheima pilosaria* on *Quercus* (UMO), *Operophtera brumata* on *Quercus* (UMO). Noctuidae – *Orthosia cerasi* (NHM).

British distribution: S.England, Midlands and Wales.
Habitat: deciduous woodland (Herting, 1960).
Flight period: late May to early June (at least 50 records).

Genus **Bessa**

B. parallela. Solitary, usually pupating in the host cocoon.

Hosts: chiefly concealed microlepidoptera larvae on trees and shrubs, commonly *Yponomeuta* spp (Yponomeutidae). In Britain: Drepanidae − *Cilix glaucata* (NMW). Noctuidae − *Nycteola revayana* (2 − NHM), *Orthosia miniosa* (NHM). Tortricidae − *Cacoecimorpha pronubana* (NHM). Yponomeutidae − *Yponomeuta padella* (4 − inc. 2 on *Crataegus*).

British distribution: S.England and Midlands (Worcs).

Flight period: late May to early September (18 records).

B. selecta. Overwinters as a first-instar larva in the host, pupating in the host cocoon or in the soil. Oviposition behaviour in Barfoot (1957).

Hosts: chiefly Tenthredinidae sawfly larvae (Hymenoptera: Symphyta) on shrubs and trees. Also a few records from Diprionidae sawflies and a weevil. In Britain: Curculionidae (Coleoptera) − *Anthonomus pomorum* (NHM). Tenthredinidae − *?Priophorus pallipes* (NHM), *Cladius pectinicornis* (NHM), *Nematus pavidus* (UMO), *?N.ribesii* on Gooseberry (Barfoot, 1957), *Hemichroa crocea* (NHM), *Pristiphora moesta* (NHM). Also from an unidentified host on *Salix* (UMO) and an unidentified sawfly on *Betula* (UMO).

British distribution: S.England, Midlands, Wales (Glamorgan) and N.England (York*).

Flight period: 2 periods − late May to early July and early August to early September (at least 50 records).

Tribe **Blondeliini**

This varied tribe contains species which lay unincubated eggs on hosts; species which lay incubated eggs on hosts; and species which pierce the host integument and insert (incubated) eggs directly into the host.

Genus **Belida**

B. angelicae. Pupates in the host cocoon.

Hosts: in Europe the sawflies *Arge nigripes* (1) and *A.berberidis* (4) (Hymenoptera: Symphyta: Argidae).

British distribution: S.England (Devils Ditch in Cambs 24 and 27.vi.1936 UMO).

Flight period: in Europe June to August (Herting, 1960).

Genus **Meigenia**

The species *dorsalis* and *mutabilis* can only be separated using the male genitalia. Literature records should therefore be treated with caution.

M. dorsalis

Hosts: Chrysomelidae larvae (Coleoptera). In Europe: *Phytodecta decemnotata* (2), *Chrysomela populi* (1), *C.tremula* (1), *Chrysolina varians* (2), the non-British *C.americana* (L.) (1) and the non-British *Crioceris quatuordecimpunctata* (Scop.) (1).

British distribution: S.England, Midlands (Hereford and Salop) and Wales.

Flight period: late May to early September, 2 generations peaking in early June and early August (at least 50 records).

M. mutabilis. Lays unincubated eggs on the older larval stages of its host. Solitary, apparently killing supernumerary individuals with their mouthparts. Pupates in the empty integument of the host, which usually survives long enough to prepare its pupation cell in the soil or to spin its cocoon.

Hosts: larvae of Chrysomelidae (Coleoptera) and *Athalia*, mostly *A.rosae* (Hymenoptera: Symphyta: Tenthredinidae). In oviposition experiments found to prefer *A.rosae* to Chrysomelidae larvae (Pschorn-Walcher, 1969). Also a single questionable record from *Hypera rumicis* (Coleoptera: Curculionidae). In Britain: Chrysomelidae (Coleoptera) – *Gastrophysa viridula* (UMO), *Phaedon cochleariae* (NHM), *Phaedon tumidulus* (NMW), *Phytodecta olivacea* (NHM). Also recorded from unidentified Curculionidae (Coleoptera) on *Chaerophyllum* (NMW). Pyralidae (Lepidoptera) – *Pleuroptya ruralis* on *Urtica* (Hamm, 1942). Tenthredinidae (Hymenoptera: Symphyta) – *Athalia ?liberta* (NHM), *Athalia cornubiae* on *Sedum album* (UMO – also in Gradwell, 1957).

British distribution: England, Wales and N.Scotland.

Habitat: Emden (1954) gives woodland but this is unlikely as most of its British hosts feed on herbs.

Flight period: late May to early September, 2 peaks – late May and late July (at least 50 records).

M. majuscula

Hosts: in the Soviet Union a record from *Chrysolina montana* Gebler (Coleoptera: Chrysomelidae).

British distribution: S.England (Felton Common in Somerset 28.viii.1937*; Malling Hill 27.x.1866 and Ranscombe 11.x.1867 both near Lewes in Sussex UMO; Deal Golf Course in Kent 7.viii.1950 NHM). Also a probably erroneous record from N.Scotland (The Mound in Sutherland, 1900*).

Habitat: British localities are coastal grassland and downland (Falk, in press). Not associated with these habitats in Europe, at northern edge of its range (Tschorsnig, pers. comm.)

Genus **Zaira**

Z. cinerea. Lays a partially incubated egg on the underside of the host abdomen. Overwinters as a second-instar larva within the host, pupating in the empty host abdomen. Solitary or gregarious depending on the host species.

Hosts: adult Carabidae (Coleoptera). In Britain: *Carabus violaceus* (NHM), *C.monilis* (Ellis, 1926) and *Pterostichus niger* (P. Chandler's collection).

British distribution: S.England, Midlands (Hereford), Wales (Glamorgan and Merioneth), N.Scotland (Ballachulish in Argyll) and Ireland (Mayo).

Flight period: early June to early August (42 records). In Europe 1 generation per year (Herting, 1960).

Genus **Gastrolepta**

G. anthracina
Hosts: in Germany a record from *Lagria hirta* (Col.: Tenebrionidae).

British distribution: S.England (Lizard Peninsula; Rame and Boscastle in Cornwall; Torcross in Devon*; Eype in Dorset; Laughton in Sussex*; Deal Golf Course in Kent).

Habitat: localities mainly coastal (Stubbs, pers. comm.)

Flight period: 2 periods – late May/early June and late July/early August (8 records), similar pattern in Germany (Tschorsnig, 1989).

Genus **Medina**

Solitary parasitoids of adult Chrysomelidae and Coccinellidae (Coleoptera). The female uses her specialised terminalia to push an unincubated egg under the host elytra. The species *luctuosa* and *separata* can only be separated using the terminalia and the latter has only recently been recognised as occurring in Britain (Ford, 1989). Literature records should therefore be treated with caution.

M. collaris. Develops in the host abdomen, overwintering as a larva within the host and leaving to pupate in the soil.

Hosts: in Britain recorded as a regular parasitoid of *Lochmaea suturalis* (Cameron *et al.*, 1944; Waloff, 1987). In Europe single records from *Lochmaea caprea* and the non-British *Galerucella luteola* (Müll.) (all Chrysomelidae). Also a record in Japan from *Epilachna vigintioctomaculata* Motsch. (Coccinellidae).

British distribution: England, Wales, S.Scotland (Dumfries), N.Scotland and Ireland (Cork).

Habitat: heathland and coastal wastes (Emden, 1954).

Flight period: late May to late August, 2 generations peaking in early June and early August (at least 50 records). The spring generation attacks overwintered adults of the univoltine *L.suturalis*, the autumn generation attacks the immature brood.

M. luctuosa. Usually solitary but occasionally 2 may complete their development in the same host. Overwinters as a second-instar larva in the host abdomen, leaving to pupate in the ground.

Hosts: in Europe reliably recorded from a number of species of *Altica*. In Britain: Chrysomelidae – *Altica lythri* (4), *A.ericeti* on *Calluna vulgaris* (NHM), *A.oleracea* (Phillips, 1977). Coccinellidae – *Adalia decempunctata* (Walker, 1962 – on *Quercus*; Banks, 1956 – on *Urtica dioica*). These last 2 records probably refer to *Medina separata*.

British distribution: S.England, Midlands, Wales (Glamorgan and Monmouth*) and N.England (Lancs).

Habitat: woodland (Emden, 1954).

Flight period: late May to late August, 2 generations peaking in early June and August (at least 50 records).

M. separata
Hosts: in Europe Chrysomelidae and Coccinellidae. In Britain: Chrysomelidae – *Phyllodecta vitellinae* on *Populus ?gileadensis* (NHM). Coccinellidae – *Adalia decempunctata* (Ford, 1989; Ford & Shaw, 1991).

British distribution: S.England (Lyndhurst in Hants*; Holt Heath in Dorset*; Barton Mills in Suffolk 25.vii.1941 NHM) and N.England (Sheffield in Yorks NHM).

Flight period: in Germany 2 peaks – late May and late July (Tschorsnig, 1989).

Genus **Policheta**

P. unicolor. Leaves the host to pupate.

Hosts: adult Chrysomelidae (Coleoptera). In Britain: *Chrysolina banksi* (NHM). In Europe single records from *C.graminis, C.haemoptera,* the non-British *C.geminata* (Payk.) and from the introduced American *Leptinotarsa decemlineata.* References to *Croesus* (Hymenoptera: Symphyta) as a host are probably erroneous.

British distribution: S.England (numerous records from Cornwall and Devon; also Dorset* and Hants*), Wales (Glamorgan) and Ireland (Cork).

Habitat: localities indicate coastal habitat, especially beaches (Falk, in press).

Flight period: late May to early October, 2 generations peaking late June and early September (42 records).

Genus **Leiophora**

L. innoxia. The female has modified terminalia indicating that the (incubated) egg may be laid in the segmental folds of the host.

Hosts: in Britain: an adult *Phyllodecta vitellinae* (Coleoptera: Chrysomelidae) (P.Chandler's collection), also a questionable record from *Tetrix tenuicornis* (Orthoptera: Tetrigidae) (NHM). In Europe unconfirmed single records from adult *Agelastica alni, Altica oleracea* and *A.brevicollis.* (all Coleoptera: Chrysomelidae), *Diprion* sp. (Hymenoptera: Symphyta) and the non-British *Tetrix bipunctata* (L.) (Orthoptera: Tetrigidae).

British distribution: S.England and Midlands (Hereford and Worcs) and an unconfirmed record from N.England (Yorks*). Commoner in more southern parts of Europe (Herting, 1960).

Flight period: July and early August (20 records). In Europe 1 generation per year (Herting, 1960).

Genus **Admontia**

Parasitoids of Tipulidae larvae (Diptera). The adults prefer wet and cold areas. *A.blanda* lays incubated eggs. Its first-instar larva is very sensitive to drying out and has little mobility, indicating the egg is probably laid directly on the host in moist conditions.

A. blanda

Hosts: in Europe single records from *Tipula hortorum* and the non-British *Nephrotoma pratensis* (L.).

British distribution: S.England (Wilts, Hants, Cambs, Norfolk, Suffolk and Surrey*), Midlands (Hereford and ?Salop), Wales (Monmouth*), N.England (Westmorland) and N.Scotland.

Flight period: July and August (19 records). In Europe June to October (Herting, 1960).

A. grandicornis
Hosts: in Britain *Tipula nubeculosa* (UMM). In Europe several records from *Tipula* sp.

British distribution: S.England (Bagley Wood in Oxford), Wales (Merioneth*), N.England (Windermere in Westmorland/Lancs) and N.Scotland (Ballater in Aberdeen; Grantown in Moray; Aviemore and Lynwilg Strathspey in Inverness).

Habitat: in Europe usually found in cool montane areas with pine forest (Tschorsnig, pers. comm.).

Flight period: early June to early August (6 records). A similar pattern in Germany – peaking in June (Tschorsnig, 1989).

A. maculisquama (= *seria* of Emden (1954)).
Hosts: in Britain: *Tipula irrorata* (Chandler, 1966), *T.meigeni* and *T.lunata* (both Chiswell, 1956). In Europe single records from *Ctenophora pectinicornis* and *C.atrata*. Records however may refer to true *seria*.

British distribution: S.England (Somerset, Wilts, Hants, Surrey*, Bucks* and Kent), Midlands, Wales (Glamorgan), N.England, S.Scotland (Kirkcudbright and Edinburgh) and N.Scotland (Perth).

Flight period: late June and July (32 records).

A. seria (= *decorata* of Emden (1954)).
Hosts: in Britain: *Ctenophora bimaculata* (NHM) and *Tipula flavolineata* in a rotting birch stump (Chandler, 1976). In Europe: *Tipula flavolineata* (1), *T.irrorata* (1) and *Ctenophora bimaculata* (3).

British distribution: S.England (Blaise Woods and Coombe Dingle in Gloucs; Mark Ash in New Forest in Hants; Ashstead woods in Surrey*; Buckhurst Hill* and Epping Forest in Essex).

Habitat: localities indicate ancient broadleaved woodland (Falk, in press).

Flight period: scattered records from June to September.

Genus **Oswaldia**

O. muscaria. Incubated eggs are laid on the host, the larva leaving it to pupate and overwinter in the soil.

Hosts: Lepidoptera larvae (Geometridae, Noctuidae and a single Thyatiridae) which are present on trees and shrubs early in the season. No British records.

British distribution: England and Wales.

Habitat: oak woodland (Herting, 1960).

Flight period: late April to late June (at least 50 records). 1 generation per year.

Genus **Hemimacquartia**

H. paradoxa
Hosts: unknown.

British distribution: N.Scotland (near Grantown in Moray, 19.vi.1943 NHM; Aviemore 26.v.1934 UMO and Glen Affric vi.1954 UMM in Inverness). A further 5 European records in May.

Genus Ligeria

L. angusticornis. Lays incubated eggs.

Hosts: all records are from Pterophoridae larvae (Lepidoptera). In Britain: *Leioptilus tephradactyla* (NHM) and *Pterophorus pentadactyla* (NHM). In Europe: *Pterophorus lithodactylus* (2), *P.pentadactyla* (1) and the non-British *P.nephelodactyla* (von Eversmann) (1).

British distribution: England and Wales (Pembroke and Glamorgan).

Flight period: early June to early August (at least 50 records).

Genus Blondelia

B. nigripes. Female abdomen and reproductive strategy similar to *Compsilura concinnata* (see below). Immature stages and biology in Dowden (1933). Overwinters in the peritrophic membrane of the mid-gut of the host, usually as a second-instar larva. Usually emerges from the host to pupate in the soil.

Hosts: a wide range of Lepidoptera and (more rarely) sawfly larvae (Hymenoptera: Symphyta). Rarely attacking hairy caterpillars (Dowden, 1933). No British host records.

British distribution: England, Wales and Ireland (near Flagstaff in Louth).

Habitat: woodland (Emden, 1954).

Flight period: late May to late August (at least 50 records). In Germany peaking May and late July/early August (Tschorsnig, 1989).

Genus Compsilura

C. concinnata. The female inserts an incubated egg directly into the host, penetrating its integument using a piercing structure developed from sternite 7. Rows of short teeth on the ventral edge of tergites 3 and 4 probably act in opposition to the piercer (see fig.326). The larva migrates to the peritrophic membrane of the host's mid-gut. Overwinters as a larva within the host, pupating in the soil or host cocoon. May be gregarious.

Hosts: larval Lepidoptera on a wide variety of plants, also a few records from sawfly larvae (Hymenoptera: Symphyta). Unlike *Blondelia nigripes*, also recorded from species with hairy caterpillars. In Britain: Arctiidae – *Arctia caja* (NHM), *Diaphora mendica* (NHM), *Spilosoma lutea* (NHM), *Spilosoma lubricipeda* (2), *Tyria jacobaeae* (NHM). Geometridae – *Abraxas grossulariata* (3), *Apocheima pilosaria* (UMO), *Lycia hirtaria* (NHM). Lasiocampidae – *Eriogaster lanestris* (NHM), *Malacosoma neustria* (8), *Poecilocampa populi* (UMO). Lymantriidae – *Leucoma salicis* (5), *Orgyia antiqua* (2 – NHM), *Calliteara pudibunda* (4), *Euproctis similis* (13), *E.chrysorrhoea* (3). Noctuidae – *Acronicta megacephala* (3 – inc. 1 on *Populus nigra*), *A.rumicis* (4), *A.tridens* on *Sorbus aucuparia* (Shaw, 1979), *Acronicta psi* (12 – inc. 1 on *Sorbus aucuparia*), *A.aceris* (22 – inc. 2 on *Acer*), *A.alni* (3 – inc. 1 on *Betula*), *Archanara dissoluta* (Hamm, 1942), *Colocasia coryli* (NHM), *Cucullia gnaphalii* (Hammond & Smith, 1957), *C.asteris* (4), *C.verbasci* (5), *Euplexia lucipara* (NHM), *Hadena bicruris* on *Lychnis dioica* (Hammond & Smith, 1953), *Hydraecia micacea* in stem of *Solanum tuberosum* (NHM), *Mamestra brassicae* (2 – inc. 1 in *Brassica oleracea*), *Moma alpium* (NHM), *Nonagria typhae* (UMO), Plusiinae sp. (Ford & Shaw, 1991), *Polymixis flavicincta* (NHM). Notodontidae – *Cerura vinula* (5), *Diloba caeruleocephala* (3),

Furcula bifida on *Populus tremula* (Hammond & Smith, 1957), *Phalera bucephala* (5). Nymphalidae – *Aglais urticae* (17), *Inachis io* (6), *Polygonia c-album* on *Ulmus* (NHM). Pieridae – *Gonepteryx rhamni* (2), *Pieris rapae* (Ford & Shaw, 1991; Hammond & Smith, 1953 – on *Brassica oleracea*), *P.brassicae* (6). Satyridae – *Lasiommata megera* (NHM). Sphingidae – *Deilephila elpenor* (NHM), *Laothoe populi* (2), *Smerinthus ocellata* (4), *Sphinx ligustri* (4). Yponomeutidae – *?Yponomeuta padella* (NHM – probably erroneous as too small). Tenthredinidae (Hymenoptera: Symphyta) – *Priophorus pallipes* (NHM), also 2 records from unidentified sawflies (inc. 1 on *Salix*).

British distribution: England and S.Scotland (Selkirk).

Habitat: gardens and wastes (Emden, 1954). In museum collections reared specimens are commoner than ones caught as adults.

Flight period: early June to late August (at least 50 records).

Genus **Vibrissina**

V. debilitata. Female abdomen similar to *Compsilura concinnata*.

Hosts: unknown. The other Palaearctic *Vibrissina* species is a parasitoid of sawfly larvae (Hymenoptera: Symphyta).

British distribution: S.England, (southern) Midlands.

Flight period: July and August (28 records – including 2 in April/May).

Tribe Winthemiini

Lays unincubated eggs on the host using a long extensible ovipositor.

Genus **Rhaphiochaeta**

R. breviseta

Hosts: unknown

British distribution: Midlands (Fiskerton in Notts* 14.vi.1919; Ross-on-Wye in Hereford* 18.v.1931). Source Wainwright (1928). A further 3 European records in May. Also rare in Europe (Herting, 1960).

Genus **Smidtia**

S. conspersa

Hosts: Geometridae larvae (Lepidoptera) present on trees and shrubs early in the year. In Britain: Geometridae – *Epirrita dilutata* (2 – inc. 1 on Quercus), *Epirrita* sp. (2 – inc. 1 on *Quercus*), *Operophtera fagata* (NHM).

British distribution: England, Wales (Caernarvon* and Merioneth), S.Scotland (Kirkcudbright*) and N.Scotland (Perth).

Habitat: deciduous woodland (Herting, 1960).

Flight period: May and early June (at least 50 records).

Genus **Timavia**

T. amoena

Hosts: in Europe scattered records from a number of Lepidoptera families on trees and shrubs, including several from *Panolis flammea* (Noctuidae). In Britain recorded ovipositing on *Orthosia incerta* (Noctuidae) (UMO).

British distribution: S.England, Midlands (Warwick) and N.England.

Habitat: pine woods (Herting, 1960).

Flight period: May and June (22 records).

Genus **Winthemia**

The species *cruentata* has only recently been recognised as occurring in Britain. Literature records for this genus should therefore be treated with caution.

W. cruentata. Gregarious, a maximum of 74 larvae being found in a single host larva (usually between 10 and 30 successfully developing).

Hosts: predominantly *Sphinx ligustri* larvae (Lepidoptera: Sphingidae). In Britain 2 records from this species (NHM). In Europe also a few other Lepidoptera, mostly Sphingidae.

British distribution: S.England (Shell Chemicals Ltd. in London NHM) plus another from an unknown locality in England (NHM).

Flight period: in Europe August, 1 generation per year (Herting, 1960).

W. quadripustulata. May be gregarious in larger hosts, leaving the host to pupate in the ground, overwintering as puparia.

Hosts: chiefly *Cucullia* larvae (Lepidoptera: Noctuidae) but also other Lepidoptera. In Britain: Arctiidae – *Tyria jacobaeae* (NHM). Noctuidae – *Cucullia verbasci* (8), *Gortyna flavago* on *Arctium lappa* (Hammond & Smith, 1955). Notodontidae – *Diloba caeruleocephala* (NHM). Nymphalidae – *Inachis io* (NHM). Sphingidae – *Sphinx ligustri* (Parmenter, 1953. Possibly refers to *cruentata*).

British distribution: England, Wales (Denbigh and Glamorgan) and N.Scotland.

Flight period: late May to late August (at least 50 records). More than 1 generation per year.

W. variegata

Hosts: in Europe 2 records from *Brachionycha sphinx* (Lepidoptera: Noctuidae). In Britain 1 questionable record from *Pavonia pavonia* (Lep.: Saturniidae) on *Quercus* (UMO).

British distribution: S.England and Midlands (Hereford and Worcs).

Flight period: May and June (26 records).

Genus **Nemorilla**

N. floralis. Pupates usually in the empty host pupa. Solitary in its microlepidopteran hosts.

Hosts: chiefly microlepidoptera larvae. In Britain recorded from the following species, indicating a possible association with thistles and nettles: Arctiidae – *Spilosoma lubricipeda* (Carr, 1935). Choreutidae – *Anthophila fabriciana* (Ford & Shaw, 1991). Noctuidae – *?Orthosia gothica* (Ford & Shaw, 1991), *Autographa gamma* (2), *Hadena perplexa* on *Silene cucubalus* (Hammond & Smith, 1955), *Plusia festucae* (NHM – no locality data). Nymphalidae – *Inachis io* (2), *Polygonia c-album* on *Urtica* (NHM). Oecophoridae – *Agonopterix* sp. on *Peucedanum* (Ford & Shaw, 1991), *Agonopterix ulicetella* (NHM), *A.propinquella* (1 – NMW and BCM). Pyralidae – *Pleuroptya ruralis* (14 – inc. 1 on *Urtica dioica* and a questionable record on *Filipendula ulmaria* NHM), *?Sitochroa verticalis* (NHM), *Eurrhypara hortulata* (2). Tortricidae – *Clepsis spectrana* (2 – inc. 1 questionable), *Acleris laterana* or *schalleriana* (LM), *Archips podana* (2 – inc. 1 on *Malus*), *Cacoecimorpha pronubana* (2), *Cnephasia stephensiana* (NHM), *Spilonota ocellana* (NHM). Also 2 records from unidentified hosts on nettle (UMO).

British distribution: England and Wales (Glamorgan and Flint).

Flight period: early May to early September, peaking late May and late July (at least 50 records).

Tribe Eryciini

With the exception of *Aplomya confinis*, all the British species in this tribe lay incubated eggs. These are either laid directly on the host or the host is contacted by an active first-instar larva.

Genus **Aplomya**

A. confinis. Lays unincubated eggs on the host.

Hosts: Lycaenidae larvae (Lepidoptera). In Europe also single records from the non-British *Agriopis bajaria* (D.&S.) (Geometridae) and (probably erroneously) from other Lepidoptera. In Britain: Lycaenidae – *Celastrina argiolus* (4), *Cupido minimus* (NHM), *Lysandra bellargus* (2), *L.coridon* (2), *Polyommatus icarus* (NHM), *Quercusia quercus* (NHM), *Strymonidia w-album* (2).

British distribution: S.England and Ireland (Down* and Mayo). Commoner in warmer central European areas (Herting, 1960) but extending to northern Sweden, Norway and Finland (Tschorsnig, pers. comm.).

Habitat: wastes, especially near coasts (Emden, 1954).

Flight period: early June to late August (at least 50 records).

Genus **Phebellia**

P. glauca

Hosts: in Europe single records from *Acronicta auricoma, A.psi, A.tridens,* (Lepidoptera: Noctuidae), *Cimbex femoratus* and *Cimbex* sp. (Hymenoptera: Symphyta: Cimbicidae). The British record in Parmenter (1953) is a misidentification.

British distribution: S.England, Midlands (Warwick), N.England (Yorks) and N.Scotland.

Habitat: usually birch woodland (Herting, 1960).

Flight period: July and August (at least 50 records). 1 generation per year.

P. glirina

Hosts: in Europe a record from *Abia sericea* (Hymenoptera: Symphyta: Cimbicidae), an exposed feeder on herbs.

British distribution: S.England, Midlands, Wales, N.England (Westmorland), N.Scotland and Ireland (Kerry, Cork, Waterford and Mayo).

Habitat: in Europe usually found in willow and poplar wetland areas (= Auen) (Herting, 1960).

Flight period: late June to late August (at least 50 records).

P. stulta

Hosts: unknown.

British distribution: S. England (Pamber Forest, Reading in Berks 17.ix.1929 UMO), Midlands (Tarrington in Hereford 14.viii.1897 UMO; Wyre Forest in Worcs 21.viii.1901 NHM). A further 11 European records in August.

P. vicina

Hosts: unknown.

British distribution: N.Scotland (near Cannich, Strath Glas in Inverness 16.vii.1936 UMO – also in Wainwright, 1940). Species known only from the male Holotype specimen. See also below.

P. villica

Hosts: in Britain *Ptilodon capucina* (Lepidoptera: Notodontidae) (NHM). No records in Europe.

British distribution: S.England (Dorset, Norfolk and Kent) and Midlands (Warwick). Also females from S.England, Midlands and N.Scotland but these may be of *vicina* (although Herting (pers. comm.) considers this unlikely).

Flight period: late June to late August (12 – including females). In Europe July and August (Herting, 1960).

Genus **Nilea**

N. hortulana. Usually gregarious, leaving the host to pupate and overwinter in the soil.

Hosts: chiefly *Acronicta* larvae (Lepidoptera: Noctuidae) but also other Lepidoptera. In Britain: Lymantriidae – *Orgyia antiqua* (2 – inc. 1 questionable NHM), *Calliteara pudibunda* (NHM). Noctuidae – *Acronicta megacephala* on *Tilia europaea* (Hammond & Smith, 1953), *A.aceris* (2), *A.tridens* (3 – inc. 1 on *Sorbus aucuparia*), *A.rumicis* (3), *A.psi* (47 – inc. 1 on *Sorbus aucuparia* and 2 on *Crataegus*), *Agrotis segetum* on *Allium cepa* (Hammond & Smith, 1953), *Autographa gamma* (2), *Mamestra brassicae* (Carr, 1935), *Xestia ashworthii* (NHM – no locality data). Notodontidae – *Phalera bucephala* (UMO). Pieridae – *?Pieris brassicae* (NHM). Tortricidae – *Archips podana* or *operana* (Hammond & Smith, 1955), *Ptycholoma lecheana* (Hammond & Smith, 1955).

British distribution: England and Wales, S.Scotland (Edinburgh* and Kirkcudbright*), N.Scotland (Monifieth in Forfar; Culbin sandhills in Inverness/Moray; Blairgowrie in Perth) and Ireland (Cork).

Flight period: late June to late August (36 records). In collections reared specimens are commoner than ones caught as adults. In Europe 1 generation per year (Herting, 1960).

Genus **Tlephusa**

T. cincinna

Hosts: in Europe 2 questionable records from *Hyloicus pinastri* (Lepidoptera: Sphingidae). The record from *Melanchra persicariae* (Emden, 1954) is erroneous.

British distribution: S.England (Bickleigh Vale in Devon; Ilsington Wood and Combe Keynes in Dorset; Cranham in Gloucs*; Farley Mount in Hants), Midlands (Pentelow in Hereford) and Wales (Rhayader in Radnor).

Habitat: localities indicate probably woodland (Falk, in press).

Flight period: early June to early August (7 records). In Europe June to August (Herting, 1960).

Genus **Epicampocera**

E. succincta. Ecological relationship with *Pieris* in Richards (1940).

Hosts: mainly *Pieris* spp (Lepidoptera: Pieridae) but also other Lepidoptera. In Britain: Noctuidae – *Hadena bicruris* (Ford & Shaw, 1991). Pieridae – *Pieris rapae* (3 – inc. 1 on *Hesperis matronalis*), *P.napi* (2 – inc. 1 on *Alliaria petiolata*). In Europe also recorded from *Evergestis forficalis* (Pyralidae) and the non-British *Ascotis selenaria* (D.&S.) (Geometridae).

British distribution: England, Wales, N.Scotland and Ireland (Cork and Waterford).

Habitat: waysides (Emden, 1954).

Flight period: late July/August (80% of records) but some from end of May (at least 50 records). In Europe more than 1 generation per year (Herting, 1960).

Genus **Phryxe**

All 4 species in this genus have been commonly confused with each other; *magnicornis* and *vulgaris* are only reliably separable by the male genitalia. Literature records should therefore be treated with caution.

P. heraclei

Hosts: *Philudoria potatoria* larvae (Lepidoptera: Lasiocampidae) with a few records from other Lepidoptera. In Britain: Lasiocampidae – *Lasiocampa quercus* (2 – inc. 1 questionable), *Malacosoma neustria* (BCM), *Philudoria potatoria* (12).

British distribution: S.England, Midlands (Cheshire), Wales and N.England (Lancs).

Habitat: marshes (Emden, 1954). Host larvae feed on coarse grasses and reeds, often found in wet areas (South, 1961).

Flight period: late May to late August (50 records – the majority in August).

P. magnicornis

Hosts: Lepidoptera larvae, mainly *Zygaena* spp (Zygaenidae) and Geometridae larvae. In Britain: Drepanidae – *Drepana cultraria* (UMO). Geometridae – *Agriopis marginaria* on *Quercus* (UMO), *Chloroclysta concinnata* (LM), *Eupithecia intricata* (NHM), *Hydriomena furcata* (2), *Opisthograptis luteolata* (Ford & Shaw, 1991), *Peribatodes rhomboidaria* (NHM), *Selenia dentaria* (NHM), *Serraca punctinalis* on *Quercus* (UMO). Lycaenidae – *Celastrina argiolus* (Hammond & Smith, 1955), *Quercusia quercus* (3 inc. 1 on *Quercus*). Notodontidae – *Diloba caeruleocephala* (NHM). Zygaenidae – *Zygaena trifolii* (2), *Z.filipendulae* (12), *Z.lonicerae* (NHM), *Zygaena* sp. (4).

British distribution: S.England, Midlands, Wales (Caernarvon*), N.England (York), N.Scotland and Ireland (Kerry and Clare*).

Habitat: wastes (Emden, 1954).

Flight period: late May to early September (based on 15 males – many other probable records but examination of male terminalia necessary).

P. nemea. Usually gregarious.

Hosts: a wide range of Lepidoptera larvae, usually avoiding strongly haired species (Herting, 1960). In Britain: Arctiidae – *Spilosoma lubricipeda* (NHM). Geometridae – *?Alcis repandata* (Ford & Shaw, 1991), *Hydriomena impluviata* (2), *Abraxas* sp. (Ford, 1973), *Abraxas grossulariata* (30 inc. 1 on *Euonymus japonicus* and 1 on black currant), *Agriopis aurantiaria* on *Quercus* (UMO), *Apocheima pilosaria* (Ford & Shaw, 1991), *Epirrita dilutata* (3 – inc. 2 on *Quercus*), *Erannis defoliaria* (7 – inc. 1 on *Quercus*), *Geometra papilionaria* (NHM), *Operophtera brumata* (Hammond & Smith, 1955; Ford, 1976; Ford & Shaw, 1991), *Semiothisa wauaria* (3). Lasiocampidae – *Lasiocampa quercus* (2). Lycaenidae – *Quercusia quercus* (9 – inc. 6 on *Quercus*), *Strymonidia pruni* (Ford & Shaw, 1991), *S.w-album* (3). Lymantriidae – *Calliteara pudibunda* (Ford & Shaw, 1991), *Euproctis similis* (Hammond & Smith, 1953 – on *Crataegus oxyacantha*). Noctuidae – *?Amphipyra* sp. (Ford & Shaw, 1991), *?Euplexia lucipara* (Ford & Shaw, 1991), *Abrostola triplasia* (4), *Acronicta psi* (NHM), *Amphipyra tragopoginis* (UMM), *Amphipyra pyramidea* (BCM), *Autographa gamma* (4 – inc. 1 on nettle), *Conistra ligula* (NHM), *Cosmia trapezina* (UMO), *Cosmia pyralina* (UMM), *Cucullia verbasci* (3), *Discestra trifolii* (Ford, 1976), *Hadena bicruris* (NHM), *Hypena rostralis* (Hammond & Smith, 1955 – on *Humulus lupulus*), *Lacanobia oleracea* (5), *L.suasa* (1 – NMW and BCM), *Mamestra brassicae* (Ford & Shaw, 1991), *Melanchra persicariae* (2), *Phlogophora meticulosa* (9), *Polychrysia moneta* (NHM on *Delphinium*), *Polymixis flavicincta* (NHM). Nymphalidae – *Aglais urticae* (5 – inc. 2 on *Urtica dioica*), *Mellicta athalia* (NHM), *Vanessa atalanta* (Audcent, 1932; Ford & Shaw, 1991). Pieridae – *Anthocharis cardamines* (2 – Ford & Shaw, 1991), *Gonepteryx rhamni* (2). Pyralidae – *Pleuroptya ruralis* on *Urtica dioica* (Davies, 1986). Saturniidae – *Pavonia pavonia* (LM). Satyridae – *Lasiommata megera* (NHM). Sphingidae – *?Hyloicus pinastri* (NHM). Thyatiridae – *Thyatira batis* (2). Yponomeutidae – *Yponomeuta* sp. (NHM). Zygaenidae – *Zygaena filipendulae* (NHM). In Europe also 2 records from sawfly larvae (Hymenoptera: Symphyta). A record from the earwig *Forficula auricularia* (Dermaptera: Forficulidae) (Hamm, 1942) is probably erroneous.

British distribution: England, Wales, S.Scotland (Kirkcudbright*), N.Scotland and Ireland (Waterford, Clare*, Wicklow, Down* and Dublin).

Habitat: usually deciduous woodland and scrub (Herting, 1960).

Flight period: early May to early October, peaking May (at least 50 records). More than 1 generation per year.

P. vulgaris. Ecological relationship with *Pieris* in Richards (1940). Usually gregarious, overwintering within the host as a third-instar larva or leaving to pupate in the ground.

Hosts: a wide range of Lepidoptera larvae, usually on herbs with approximately half the records from Pieridae and Nymphalidae. In Britain: Arctiidae – *Callimorpha dominula* (NMW), *Phragmatobia fuliginosa* (UMO). Geometridae – *?Semiothisa notata* (NHM), *Eulithis pyraliata* (NHM), *Lycia zonaria* (NHM), *Pseudoterpna pruinata* (NHM). Hesperiidae – *Thymelicus lineola* (Ford & Shaw, 1991), *T.sylvestris*, *T.acteon* (NHM). Lasiocampidae – *Philudoria potatoria* (3 – genitalia not examined). Lycaenidae – *Lycaena dispar* (NHM), *Lysandra coridon* (2). Noctuidae – *?Agrochola lychnidis* (NHM), *?Dypterygia scabriuscula* (NHM), *Acronicta ?aceris* (NHM), *Anarta myrtilli* (NHM), *Autographa gamma* (2 – inc. 1 on nettle), *Cucullia verbasci* (BCM), *Eremobia ochroleuca* (Ford & Shaw, 1991), *Euplexia lucipara* (NHM), *Hadena bicruris* (Ford & Shaw, 1991), *Hecatera bicolorata* (2 – inc. 1 on *Heracleum*), *Lacanobia oleracea* (UMO), *Mamestra brassicae* (NHM), *Phlogophora meticulosa* (NHM), *Plusia festucae* (2), *Polymixis flavicincta* (UMO). Notodontidae – *Diloba caeruleocephala* on *Crataegus* (Hammond & Smith, 1953). Nymphalidae – *Aglais urticae* (14), *Cynthia cardui* (BCM),

Inachis io (2), *Nymphalis polychloros* (NHM). Pieridae – *Anthocharis cardamines* (12 – inc. 1 on *Alliaria petiolata*), *Aporia crataegi* (NHM – now extinct in Britain), *Pieris rapae* (17), *Pieris napi* (3), *Pieris brassicae* (UMO). Saturniidae – *Pavonia pavonia* (NHM). Sphingidae – *Hyloicus pinastri* (2 – inc. 1 questionable), *Deilephila porcellus* (NHM). Zygaenidae – *Zygaena* sp. (UMO). Also recorded from *Neodiprion sertifer* (Hymenoptera: Symphyta: Diprionidae) on *Pinus sylvestris* (NHM).

British distribution: England, Wales, N.Scotland and Ireland (Cork, Clare, Dublin and Kerry*).

Habitat: open (Herting, 1960; Ford & Shaw, 1991; Belshaw, 1992).

Flight period: early May to late September, peaking August (at least 50 records). More than 1 generation per year.

Genus **Bactromyia**

B. aurulenta

Hosts: Lepidoptera larvae on trees and shrubs; in Europe one third of the records from *Yponomeuta* spp (Yponomeutidae). In Britain: Geometridae – *Drepana binaria* (Imperial College at Silwood Park) and either *Apocheima pilosaria* or *Philereme transversata* (Ford & Shaw, 1991). Lycaenidae – *Quercusia quercus* (NHM), *Strymonidia w-album* (UMO). Lymantriidae – *Calliteara pudibunda* (Ford & Shaw, 1991). Noctuidae – *Bena prasinana* (2 – on *Fagus* and *Quercus*), *Nycteola revayana* (NHM), *Orthosia gracilis* (NHM).

British distribution: S.England and Midlands (Hereford and Worcs*).

Flight period: late May to early September (8 records). In Europe 2 generations per year (Herting, 1960).

Genus **Pseudoperichaeta**

P. nigrolineata. Lays the egg in the vicinity of the host which is then contacted by the mobile first-instar larva. The autumn generation overwinters as a second-instar larva within the host. Pupation occurs at the host's feeding location next to its remains (the host may survive long enough to pupate). Usually solitary. Aspects of biology in Plantevin *et al*. (1986) and Ramadhane *et al*. (1987).

Hosts: approximately 90% of the records from larvae of Tortricidae and Pyralidae (concealed Lepidoptera). In Britain: Geometridae – *Entephria caesiata* (UMM). Hesperiidae – *Thymelicus sylvestris* or *lineola* (NHM). Oecophoridae – *Agonopterix conterminella* (UMM). Pyralidae – *Eurrhypara hortulata* (4), *Mutuuraia terrealis* (Ford & Shaw, 1991), *Pleuroptya ruralis* (Ford, 1976). Tortricidae – ?*Hedya dimidioalba* (NHM), *Acleris rhombana* (NHM), *Acleris logiana* (NHM), *A.rhombana* (NHM), *A.rosana* (4), *Archips podana* (NHM), *A.xylosteana* on *Quercus* (UMO), *Cacoecimorpha pronubana* (3), *Cnephasia stephensiana* (Ford & Shaw, 1991), *Cydia pomonella* (Ford & Shaw, 1991), *Tortrix viridana* (NHM). Also recorded from unidentified Tortricidae on *Crataegus*, red currant and *Quercus*. Yponomeutidae – *Scythropia crataegella* (Audcent, 1932).

British distribution: England, Wales (Glamorgan) and S.Scotland (Midlothian).

Habitat: woodland (Emden, 1954).

Flight period: 2 periods – May/early June and early July to late August (at least 50 records).

Genus **Lydella**

Lays incubated eggs on the host plant, the host then contacted by the mobile first-instar larva.

L. grisescens
Hosts: in Britain: *Peridea anceps* (Lepidoptera: Notodontidae) (NHM) and either *Dypterygia scabriuscula* (Lep.: Noctuidae) or *Hyloicus pinastri* (Lep.: Sphingidae) (NHM). In Europe a questionable record from *Archanara neurica* (Lep.: Noctuidae).

British distribution: England and Wales.

Flight period: early June to early September (at least 50 records).

L. stabulans
Hosts: stem-boring Noctuidae larvae (Lepidoptera), usually those in wetland plants such as *Juncus, Typha* and *Phragmites*. Records here from other species are questionable. In Britain: Lymantriidae – *Lymantria monacha* (NHM). Noctuidae – *Archanara dissoluta* (NMW), *A.geminipuncta* (NHM), *Arenostola phragmitidis* (UMO), *Coenobia rufa* (NHM), *Hydraecia micacea* (2 – inc. 1 in *Rumex* stem), *Lacanobia oleracea* (NHM), *Nonagria typhae* (Ford & Shaw, 1991), *Oligia strigilis* (NHM), *Photedes minima* in *Deschampsia caespitosa* (NHM). Nymphalidae – *Nymphalis polychloros* (NHM). Sphingidae – *Acherontia atropos* (NHM). Also recorded from an unidentified host in a thistle stem (LM).

British distribution: England, Wales, N.Scotland and Ireland (Wicklow, Dublin, Down* and Mayo).

Flight period: late May to early September (at least 50 records). More than 1 generation per year.

Genus **Cadurciella**

C. tritaeniata. Lays incubated eggs, its long ovipositor indicating that these are laid directly on the host.

Hosts: specific to *Callophrys rubi* larvae (Lepidoptera: Lycaenidae). In Britain 9 records from this species, including 1 on *Cornus* sp.

British distribution: S.England, Midlands (Hereford and Warwick), Wales (Radnor) and N.Scotland (Moray and Inverness).

Flight period: June and July (14 records). In Europe mid-May to the beginning of July (Herting, 1960).

Genus **Drino**

D. lota. Lays incubated eggs on the host, up to 27 adults reared from a single individual. Overwinters as a puparium in the ground.

Hosts: chiefly *Deilephila elpenor* larvae (Lepidoptera: Sphingidae). In Britain 6 records from this species, including 1 on *Epilobium hirsutum*. In Europe also single records from *D.porcellus, Laothoe populi* (Sphingidae) and the non-British *Aglia tau* (L.) (Lep.: Saturniidae).

British Distribution: S.England, Midlands (Hereford* and Derby), S.Scotland (Dunbarton*), N.Scotland (Inverness*) and Ireland (Mayo*).

Flight period: late June to late August (20 records). In Europe mid-June to early September, 1 generation per year (Herting, 1960).

Genus **Huebneria**

H. affinis

Hosts: in Europe numerous records from Arctiidae larvae (Lepidoptera), mostly *Arctia caja, A.villica, Parasemia plantaginis, Phragmatobia fuliginosa* and the non-British *Ammobiota festiva* (Hufnagel). Also scattered records from a range of other Lepidoptera, usually ones with hairy caterpillars. In Britain 1 record from *Arctia caja* (UMO).

British distribution: S.England (Folkestone 1866 and Deal 18.vi.1921 in Kent both UMO).

Flight period: in Europe May to September (Herting, 1960).

Genus **Carcelia**

Lays incubated eggs on hairy Lepidoptera caterpillars. These eggs have a characteristic stalk with which they are attached to a hair of the host. Several species have only recently been recognised as British (following Herting, 1977) and may be commoner and more widely distributed than is indicated here. Literature records should therefore be treated with caution (*bombylans* probably not British).

C. atricosta

Hosts: in Britain the following tree and shrub-feeding species: Lasiocampidae – *Malacosoma neustria* (2). Lymantriidae – *Orgyia antiqua* (4). Noctuidae – *Acronicta psi* (NHM). In Europe a further 3 records from *O.antiqua*.

British distribution: S.England (Cornwall, Hants, Berks, Surrey and London), Wales (Caernarvon), N.England (Lancs) and N.Scotland (Watten and Wick in Caithness).

Flight period: late June to early August (8 records).

C. gnava. Up to 4 individuals developing in a single host, pupating in the host cocoon or in the soil. Overwinters as a larva in the host larva or cocoon.

Hosts: in Britain: Arctiidae – *Arctia caja* (NMD). Lasiocampidae – *Malacosoma neustria* (9). Lymantriidae – *Calliteara pudibunda* (9 – inc. 1 on *Quercus*). Notodontidae – *Phalera bucephala* (NHM). Thyatiridae – *Polyploca ridens* (BCM). In Europe three-quarters of the records from *Leucoma salicis* (Lymantriidae), *M.neustria* and *C.pudibunda*, the first 2 species being attacked by the spring generation of the tachinid and last 1 by the summer generation. Also scattered records from other Lepidoptera.

Distribution: S.England, Midlands, Wales and N.England (Lancs*) and Ireland (Mayo*, Dublin – NMD, and Kerry – M. Speight).

Habitat: woodland (Emden, 1954). Its usual hosts all feed on trees and shrubs.

Flight period: early May to early July (22 records). In Europe 2 generations, May/June and July/August (Herting, 1960).

C. lucorum. Up to 5 individuals developing in a single host. Overwintering as a larva within the host, leaving it in the spring to pupate.

Hosts: mainly Arctiidae larvae (Lepidoptera) but also other Lepidoptera. In Britain: Arctiidae – *Arctia villica* (5), *A.caja* (35 – inc. 1 on *Lupinus polyphyllus*), *Parasemia plantaginis* (2), *Phragmatobia fuliginosa* (9), *Spilosoma lubricepeda* or *lutea* (NHM). Lasiocampidae – *Philudoria potatoria* (UMM).

British distribution: England, Wales (Caernarvon and Flint) and Ireland (Fermanagh* and Sligo − P.Chandler)

Flight period: late April to early September, peaking in May and August (at least 50 records).

C. puberula

Hosts: in Europe a record from *Lymantria monacha* (Lepidoptera: Lymantriidae).

British distribution: S.England (Winsford in Somerset 4.v.1972; Silwood Park in Berks 13-15.v. and 8-10.vii.1989; Ranmore in Surrey 15.v.1955 all NHM).

Habitat: in Europe males observed hovering in woodland clearings (Herting, 1960).

Flight period: in Europe most common in May and June (Herting, 1960).

C. rasa. Usually 2-4 individuals develop in a single host, pupating in the host cocoon.

Hosts: almost exclusively Lymantriidae (Lepidoptera). In Britain *Orgyia antiqua* (3) (Lymantriidae).

British distribution: S.England (Berks, Hants, Surrey, Sussex and Essex) and Midlands (Hereford and Worcs).

Flight period: late May to late August (8 British and 7 German records).

C. tibialis

Host: no reliable records. Reported in Britain from *Arctia caja* (Lepidoptera: Arctiidae) (Carr, 1935). Reported in Europe from *Calliteara pudibunda* and *Dicallomera fascelina* (both Lep.: Lymantriidae).

British distribution: S.England and Midlands (Notts*).

Flight period: late May and June (at least 50 records).

Genus **Senometopia**

Biology as in *Carcelia* except this genus tends to attack Lepidoptera caterpillars without long hairs. All 3 species are very difficult to separate.

S. excisa

Hosts: in Europe a range of Lepidoptera larvae. In Britain *Abraxas sylvata* (Geometridae) (BCM), although identification of the Tachinidae is not certain.

British distribution: S.England (Lynton in Devon 29.vii.1895 NHM; ?Painswick in Gloucs BCM).

Flight period: in Europe June to August (Herting, 1960).

S. intermedia. Solitary, pupating within the host pupa.

Hosts: in Britain *Abraxas sylvata* on *Ulmus glabra* (Lepidoptera: Geometridae) (UMO − also in Smith, 1961). In Europe recorded from the closely related *Lomaspilis marginata* (Geometridae).

British distribution: S. England (Wick in Hants; Failand in Somerset; Wrekin in Salop) and Midlands (Hereford)

Flight period: late June to early September (5 records). In Europe end of July and August (Herting, 1960).

S. pollinosa. Biology described in detail in Herrebout (1966 and 1969). Overwinters as a larva in the sub-alar cavity of the host, leaving in May to pupate in the ground.

Hosts: *Bupalus piniaria* and, to a lesser extent, *Semiothisa liturata* and *Thera obeliscata*, (all Lepidoptera: Geometridae feeding on *Pinus*). Also occasional records from other Lepidoptera on *Pinus*. In Britain 3 records from *B.piniaria*.

British distribution: England.

Habitat: pine woodland, but also found in neighbouring oak copses (Herting, 1960). These may be females in their preoviposition period (see Herrebout & Veer, 1969).

Flight period: July and August (20 records). 1 generation per year, synchronised with *B.piniaria*.

Genus **Thecocarcelia**

T. acutangulata. First recorded as British in Wyatt (1986).

Hosts: in Europe a record from *Thymelicus lineola* (Lepidoptera: Hesperiidae). Also questionable records in Finland from *Clostera pigra* and *C.curtula* (Lep.: Notodontidae).

British distribution: S.England (Hadleigh Wood in Essex 21-22.viii.1954 NHM – also in Wyatt, 1986). Rarer in northern Europe (Tschorsnig, pers. comm.).

Genus **Erycia**

E. furibunda

Hosts: in Britain: *Eurodryas aurinia* (Lepidoptera: Nymphalidae) (6) and *Lasiocampa trifolii* (Lep.: Lasiocampidae) (NHM). In Europe a record from the non-British *Hypodryas maturna* (L.) (Lep.: Nymphalidae).

British distribution: S.England (Cornwall, Devon, Dorset and Wilts), Midlands (Warwick*) and Wales (Glamorgan).

Habitat: localities and habitat preference of its usual host indicate marshland and damp unimproved meadows (Falk, in press).

Flight period: late June to early September (7 records).

Genus **Xylotachina**

X. diluta. The female possesses a long ovipositor but the host is presumably contacted by the first-instar larva. Pupation occurs in the host gallery.

Hosts: specific to wood-boring *Cossus cossus* larvae (Lepidoptera: Cossidae). In Britain 1 record from this host (NHM – in *Quercus*).

British distribution: S.England (Lyndhurst Road in Hants 1.vii.1897 UMO; Ascot in Berks NHM). In Europe also rare (Herting, 1960).

Flight period: in Europe July (Herting, 1960).

Genus **Townsendiellomyia**

T. nidicola. Biology in Muesebeck (1922). Eggs are laid on the ventral surface of the host between the legs, usually on the thorax. Solitary, overwintering as a first-instar larva in a cyst attached to the host oesophagous. Pupating within the host integument after it has formed its cocoon. First recorded as British in Wyatt & Sterling (1988).

Hosts: specific to *Euproctis chrysorrhoea* (Lepidoptera: Lymantriidae). In Britain 2 records from this host.

British distribution: S.England (Farlington Marshes in Hants, 1984* – see Wyatt & Sterling (1988); London S.E.11 1984 NHM). Rarer in northern Europe (Tschorsnig, pers. comm.).

Flight period: in Europe mid-July to mid-August (Herting, 1960).

Tribe Goniini

Species in this tribe lay small incubated eggs on the host food plant, which are ingested by the host and hatch in the host gut.

Genus **Platymya**

P. fimbriata

Hosts: no confirmed host records. Those in Emden (1954) are considered probably erroneous by Herting (1960). In N.America a species of *Platymya* has been recorded from *Crambus* (Lepidoptera: Pyralidae).

British distribution: S.England, Midlands, Wales (Glamorgan and Merioneth), N.England (Lancs and Yorks), N.Scotland (Inverness and Ross) and Ireland (Kerry).

Flight period: late May to early September (at least 50 records).

Genus **Eumea**

E. linearicornis

Hosts: in Europe: Noctuidae and Tortricidae larvae (Lepidoptera), the majority feeding on trees and shrubs. In Britain: Noctuidae – *Cosmia pyralina* on *Ulmus campestris* (Hammond & Smith, 1955), *Orthosia cerasi* on *Quercus* (UMO), *O.populeti* (BCM). Also from an unidentified Noctuidae on *Aster* (Ford & Shaw, 1991). Pyralidae – *Eurrhypara hortulata* (2). Also a record from this species in Europe. Tortricidae – *Archips ?crataegana* on *Quercus* (UMO).

British distribution: S.England, Midlands and Wales (Pembroke and Caernarvon).

Habitat: deciduous woodland (Herting, 1960)

Flight period: early May to early September (at least 50 records). In Europe 2 generations per year (Herting, 1960).

Genus **Myxexoristops**

Parasitoids of sawfly larvae (Hymenoptera: Symphyta). The two species have up until recently been confused and both included in *blondeli*. Herting (1960) describes *blondeli* as found in woodland. Overwinters as a fully developed larva within the empty host integument (Wardle, 1914).

M. blondeli

Hosts: in Britain 2 records from *Neurotoma saltuum* (Symphyta: Pamphiliidae) (NHM – 1 on *Crataegus*). In Europe single records from this host, *Mesoneura opaca* and *Pristiphora moesta* (both Symphyta: Tenthredinidae). The record from *Eriocampa ovata* (Parmenter, 1953) is a misidentification.

British distribution: S.England (Vernditch in Wilts 8.vii.1967 NHM; Burnham Beeches in Bucks NHM; Hemel Hempstead in Herts NHM).

Flight period: end of May to end of July (23 European records).

M. stolida

Hosts: in Britain: Tenthredinidae – *Eriocampa ovata* (NHM), *Pristiphora erichsonii* (3), *Croesus latipes* on *Betula* (2), *Hemichroa crocea* (NHM). All these species are tree-feeders. In Europe recorded from a number of other Tenthredinidae sawflies (Nematinae).

British distribution: S.England and N.England (Cumberland).

Flight period: early June to late August (6 British and 8 European records).

Genus **Zenillia**

Z. libatrix. May be gregarious. Overwinters as a second- (occasionally first-) instar larva in the host, pupating in the host cocoon. Biology and immature stages in Dowden (1934).

Hosts: Lepidoptera larvae, usually species which feed on trees. In Britain: Drepanidae – *Drepana cultraria* (UMO). Geometridae – *Archiearis notha* (NHM). Lasiocampidae – *Malacosoma neustria* (NHM). Lymantriidae – *Euproctis similis* (2). Noctuidae – *Hadena bicruris* (NHM), *Xanthia gilvago* (NHM).

British distribution: S.England and Midlands (Hereford*).

Flight period: early May to late September (22 records). In Europe 2 generations – May/June and August/September (Herting, 1960).

Genus **Clemelis**

C. pullata

Hosts: Pyralidae (Lepidoptera), in 2 studies found to be an important parasitoid of *Margaritia sticticalis*. Also a few records from other microlepidoptera. The only British record (*Hadena bicruris* (Parmenter, 1953)) is a misidentification.

British distribution: S.England (Headley in Surrey 16.vii.1950 NHM; Eynesford 10.vi. and 11.viii.1934 NHM and BCM and Soakham Down 6.viii.1938 UMO both in Kent; Lea Valley in Essex*). Rarer in northern Europe although range extends to northern Sweden and Finland (Tschorsnig, pers. comm.).

Flight period: in Europe 2 generations per year, May/June and August (Tschorsnig, pers. comm.).

Genus **Pales**

P. pavida

Hosts: a wide range of Lepidoptera larvae in a variety of habitats. In Britain: Arctiidae – *Arctia caja* (BCM). Geometridae – *?Erannis defoliaria* on *Quercus* (UMO), *Agriopis leucophaearia* (Ford & Shaw, 1991), *A.marginaria* on *Quercus* (3

– UMO), *A.aurantiaria* on *Quercus* (2 – UMO), *Alsophila aescularia* (2 – inc. 1 on *Quercus*), *Odontopera bidentata* (Hammond & Smith, 1955). Hesperiidae – *Thymelicus lineola* (NHM). Lasiocampidae – *Lasiocampa quercus* (UMO), *Malacosoma neustria* (7 – inc. 1 on *Crataegus oxyacantha*). Lycaenidae – *Lycaena phlaeas* (UMO). Lymantriidae – *Leucoma salicis* (Ford & Shaw, 1991), *Orgyia antiqua* (BCM), *Euproctis chrysorrhoea* (2). Noctuidae – *Heliothis armigera* (NHM), *?Agrotis segetum* (NHM), *Acronicta alni* (Ford & Shaw, 1991), *A.rumicis* (Ford & Shaw, 1991), *Agrochola litura* (Ford & Shaw, 1991), *Amphipyra pyramidea* (UMO), *Apamea crenata* (Ford, 1976), *Diarsia brunnea* (UMM), *Lacanobia oleracea* (Ford, 1976), *Mythimna straminea* (Ford & Shaw, 1991), *Mythimna* sp. (3 – Ford & Shaw, 1991), *Noctua pronuba* (Ford & Shaw, 1991), *Orthosia miniosa* (2), *O.cruda* on *Quercus* (3 – UMO), *O.munda* on *Quercus* (UMO), *O.cerasi* (9), *Orthosia* sp. (Ford & Shaw, 1991). Notodontidae – *Ptilodon capucina* (2 – Ford & Shaw, 1991). Nymphalidae – *Aglais urticae* (BCM), *Vanessa atalanta* (Ford & Shaw, 1991). Pterophoridae – *Oidaematophorus lithodactyla* (Audcent, 1932). Sphingidae – *Mimas tiliae* (NMW). Thyatiridae – *Polyploca ridens* (Ford & Shaw, 1991). Zygaenidae – *Zygaena lonicerae* (7), *Z.filipendulae* (11), *Z.trifolii* (NHM).

British distribution: England, Wales, S.Scotland (Dumfries*), N.Scotland (Culbin Sands in Inverness/Moray) and Ireland (unknown locality – SI and Clare*)

Flight period: 2 periods – late May/early June and early July to late September (at least 50 records). Possibly more than 2 generations.

Genus **Phryno**

P. vetula

Hosts: tree-feeding Lepidoptera larvae. In Britain: Geometridae – *Apocheima pilosaria* (Carr, 1935; UMO – on *Quercus*). Lasiocampidae – *Lasiocampa quercus* (Ford, 1976). Noctuidae – *Dichonia aprilina* (NMW), *Dryobotodes eremita* on *Quercus* (UMO), *Orthosia miniosa* (2). Also a questionable record of *Orthosia cerasi* on *Quercus* (UMO).

British distribution: England and Wales.

Habitat: deciduous woodland (Herting, 1960).

Flight period: late April and May (at least 50 records).

Genus **Cyzenis**

C. albicans. Solitary, pupating in the empty host pupa. See O'Hara & Cooper (1992) for references on the biology of this species.

Host: chiefly *Operophtera brumata* (Lepidoptera: Geometridae) larvae, also other tree-feeding Lepidoptera. In Britain: Geometridae – *Eupithecia pimpinellata* (BCM), *Operophtera brumata* (14 – inc. 5 on *Quercus*), *O.fagata* (Ford & Shaw, 1991). Noctuidae – *Nycteola revayana* (NHM). Yponomeutidae – *Ypsolopha vittella* on *Ulmus glabra* (Ford & Shaw, 1991).

British distribution: England, Wales, S.Scotland (Dumbarton, Edinburgh* and Kirkcudbright*), N.Scotland and Ireland (Wicklow).

Habitat: deciduous woodland and fruit orchards (Herting, 1960).

Flight period: early April to early June (at least 50 records). 1 generation per year.

Genus **Erycilla**

E. ferruginea
Hosts: in the Netherlands a record from Tipulidae larvae (Diptera) (Mesnil, 1944-1975). Also a few old and probably erroneous European records from Lepidoptera larvae.
British distribution: S.England (Devon), Midlands, Wales (Radnor), N.England and N.Scotland.
Habitat: wooded areas (Herting, 1960).
Flight period: early June to early August (21 records).

Genus **Ocytata**

O. pallipes. Eggs laid on leaves, flowers or fruit that the host has fed on the previous night. Overwinters as a second-instar larva in the host, leaving to pupate.
Hosts: the earwig *Forficula auricularia* (Dermaptera; Forficulidae), see Phillips (1983). In the Soviet Union also recorded from the non-British *F.tomis* (Kolenati). Records from Lepidoptera larvae are probably erroneous (Herting, 1960).
British distribution: England, S.Scotland (Edinburgh) and N.Scotland (Moray* and Orkney*).
Habitat: woodland and gardens (Herting, 1960).
Flight period: early June to late September, peaking July (at least 50 records). 1 generation per year.

Genus **Eurysthaea**

E. scutellaris
Hosts: chiefly microlepidoptera larvae (Yponomeutidae, Tortricidae and Pyralidae), the majority of records from *Yponomeuta*. Also a few records from Geometridae and small Noctuidae (Lep.).
British distribution: S.England (Bexley in Kent 9.vi.1903 NHM). Rarer in northern Europe (Tschorsnig, pers. comm.).
Flight period: late May to early September (21 European records).

Genus **Erynnia**

E. ocypterata. Pupates in the empty host pupa. Previously known as *Erynnia nitida* (R.-D.), this species should not be confused with the non-British *Erynniopsis antennata* Rondani which has in the past also been (erroneously) called *Erynnia nitida* (R.-D.).
Hosts: in Europe: *Sparganothis pilleriana* (Lepidoptera: Tortricidae) (4) and the non-British *Gelechia obscuripennis* (Frey) (Lep.: Gelechiidae) (1).
British distribution: S.England and N.England.
Habitat: marshes (Emden, 1954).
Flight period: 2 periods – May/June and August/September (20 records). Similar pattern in Europe (Herting, 1960)

Genus **Elodia**

E. ambulatoria

Hosts: Tineidae larvae (Lepidoptera) feeding in bracket fungi (Basidiomycotina: Aphyllophorales). In Britain 3 records from *Morophaga choragella* – 2 in *Pseudotrametes gibbosa* (Pers. ex Pers.) Bond & Sing (NHM and UMO) and 1 in an unidentified bracket fungus (NHM). Also recorded from an unidentified Lepidoptera larva in *Bjerkandera adusta* (Willd. ex Fr.) Karst (Chandler, 1976). In Europe further records from *M.choragella* (10) and *Nemapogon* spp (5).

British distribution: S.England (Gloucs*, Berks, Hants, Surrey*, London and Kent*).

Habitat: localities indicate woodland (Falk, in press).

Flight period: late May to early September (6 British and 3 European records).

E. morio. When attacking *Cydia pomonella* the eggs are laid in the entrance hole made by the host in the fruit (Kahrer, 1987). Pupates in the empty host pupa.

Hosts: concealed microlepidoptera on shrubs and (to a lesser extent) trees. In Britain: Gelechiidae – *Anacampsis populella* (2), *Gelechia sororculella* (Ford & Shaw, 1991). Oecophoridae – *Agonopterix ocellana* (BCM). Tineidae – *Tinea trinotella* (NHM). Tortricidae – *Apotomis capreana* (BENHS), *?Hedya dimidioalba* (NHM), *?Pandemis heparana* (UMO), *?Tortrix viridana* (NHM), *Acleris variegana* on *Rosa* (NHM), *A.hastiana* (2), *Cydia pomonella* (4), *C.servillana* (UMO), *Epinotia sordidana* (NHM).

British distribution: S.England, Midlands, N.England (Lancs), S.Scotland (Stirling*) and N.Scotland.

Habitat: woodland and scrub (Herting, 1960).

Flight period: early May to early August (17 records). In Europe 2 generations – May/June, and end of July to September (Herting, 1960).

Genus **Hebia**

H. flavipes

Hosts: in Britain: *Colotois pennaria* (Lepidoptera: Geometridae) on *Quercus* (3 – UMO) and *Orthosia miniosa* (Wainwright, 1905 – as *Ocytata pallipes* see Wainwright, 1928). In Europe: *C.pennaria* (4) and *Larentia* sp. (1) (Geometridae).

British distribution: S.England and Midlands.

Habitat: deciduous woodland (Herting, 1960).

Flight period: late April to early June (38 records).

Genus **Frontina**

F. laeta. Gregarious. Its main host feeds nocturnally.

Hosts: mainly *Smerinthus ocellata* larvae (Lepidoptera: Sphingidae). Five records from this species in Britain (including 2 on *Salix*) and 3 in Europe. In Europe also recorded from *Laothoe populi* (1) and *Sphinx ligustri* (2) (both Sphingidae).

British distribution: S.England (Dorset, Somerset, Hants, Surrey and Middlesex).

Habitat: most localities are either heathland or grassland (Falk, in press).

Flight period: July and August (19 records). In Europe end of July to the beginning of September (Herting, 1960).

Genus **Thelymorpha**

T. marmorata
Hosts: hairy Lepidoptera larvae in a number of families, mainly *Arctia caja* (Arctiidae) and *Malacosoma neustria* (Lasiocampidae). In Britain: Arctiidae – *Arctia caja* (2). Lasiocampidae – *Eriogaster lanestris* or *Malacosoma neustria* (NHM). Lymantriidae – *Lymantria dispar* (NHM).

British distribution: S.England, Midlands (Worcs) and Ireland (Kerry, Cork and Mayo).

Flight period: 2 periods – May/early June and late July to early September (31 records). A similar pattern in Europe (Herting, 1960).

Genus **Brachicheta**

B. strigata
Hosts: unknown. References to *Notodonta dromedarius* (Lepidoptera) as a host (Emden, 1954) are probably erroneous given its phenology (see Heath & Emmet, 1979). In Europe *B.strigata* adults are usually found near the ground (Herting, 1960).

British distribution: S.England, Midlands, Wales and N.Scotland (Aviemore in Inverness).

Habitat: localities are broadleaved woodland (Falk, in press), but also recorded from Stone Marshes (BENHS) which is open marshland (Chandler, pers. comm.).

Flight period: late March to early May (27 records).

Genus **Gonia**

G. capitata
Hosts: in Europe: larvae of *Agrotis* spp (6) and *Euxoa obelisca* (1) (both Lepidoptera: Noctuidae: Noctuinae). In Britain an old record from *Ceramica pisi* (Audcent, 1942).

British distribution: S.England (Dorset, Hants, Berks, Sussex and Kent) and Ireland (Finner sandhills, Bundoran in Donegal*).

Habitat: U.K. localities are grassland, usually near coasts, with some sites referring to chalk downland (Falk, in press).

Flight period: late July and August (31 records).

G. divisa
Hosts: in Europe an old record from *Agrotis segetum* (Lepidoptera: Noctuidae).

British distribution: S.England.

Habitat: localities indicate broadleaved woodland and heathland (Falk, in press). In Europe usually found in open woods on sand (Herting, 1960).

Flight period: late March to early May (at least 50 records).

G. ornata
Hosts: in Europe single records from *Euxoa nigricans, Colocasia coryli, Autographa gamma* (all Lepidoptera: Noctuidae), *Dasychira pudibunda* (Lep.: Lymantriidae) and in North Africa *Psilogaster loti* (Ochsenheimer) (Lep.: Lasiocampidae). Also a few questionable records from other Lepidoptera. In Britain *Ceramica pisi* (Noctuidae) (UMM).

British distribution: S.England, Wales (Merioneth and Glamorgan), N.England,

N.Scotland (Clyde Isles, Inverness* and Forfar) and Ireland (Wicklow and Mayo –
M. Speight).

Flight period: early April to early June (at least 50 records).

G. picea. Pupates in the empty host pupa, overwintering as an uneclosed adult.

Hosts: in Europe an important parasitoid of *Cerapteryx graminis*, but also single
records from *Xestia xanthographa*, the non-British *Staurophora celsia* (L.) and, in
Japan, from *Mamestra brassicae* (all Lepidoptera: Noctuidae). In Britain:
Mythimna comma (Noctuidae) (Wainwright, 1940) and *Polygonia c-album* (Lep.:
Nymphalidae) (NHM).

British distribution: S.England, Midlands, Isle of Man and Ireland (Fermanagh).

Habitat: long grass on chalk (Emden, 1954). In Europe usually found in
meadows and woodland margins (Herting, 1960).

Flight period: late March to early May (at least 50 records).

Subfamily Tachininae

Species in this subfamily lay incubated eggs, usually in the vicinity of the host which
is then contacted by the first-instar larva. The larva is either mobile or attached
posteriorly by the egg case to the leaf (see fig.307), making pendulum-like
movements with its anterior body to contact the host.

Tribe Tachinini

Genus **Tachina**

T. grossa

Hosts: mainly *Lasiocampa quercus* (7 British records) and *Macrothylacia rubi* (2
British records) (both Lepidoptera: Lasiocampidae). In Europe also other large
hairy Lepidoptera larvae: *Lasiocampa trifolii* (1), *Dendrolimus* spp (3) (Lasiocam-
pidae), *Lymantria dispar* (2) (Lep.: Lymantriidae) and *Hyloicus pinastri* (1)
(Sphingidae). Emden (1954) also lists *Euproctis chrysorrhoea* (Lymantriidae) and
Hemaris fuciformis (Lep.: Sphingidae) as hosts.

British distribution: S.England, Midlands (Warwick and Staffs), Wales, N.Eng-
land, S.Scotland (Dumfries and Kirkcudbright*), N.Scotland and Ireland (Kerry,
Waterford, Mayo, Wicklow, Galway, Louth, Kildare, Clare, Antrim, Tipperary,
Down, Dublin, Wexford* and Fermanagh*). The majority of British records are
from Cornwall, Devon, Dorset and the New Forest (Hants).

Habitat: heathland and open woodland (Herting, 1960).

Flight period: July and August (at least 50 records).

T. fera. The first-instar larva lies flat on the leaf, stimulated by vibration to take up
the host-searching posture shown in fig.307.

Hosts: Noctuidae larvae (Lepidoptera). In Britain: *Ceramica pisi* (NHM),
Cosmia trapezina on *Quercus* (UMO), *Orthosia cruda* on *Quercus* (UMO) and
O.cerasi on *Quercus* (UMO).

British distribution: S.England, Midlands, Wales, N.England (Yorks), N.Scot-
land (Rannoch in Perth; Loch Assynt in Sutherland) and Ireland (Waterford,
Kerry, Mayo, Galway and Wicklow).

Flight period: 2 periods – May/early June and late July to late September (at least
50 records).

T. lurida. Pupates in the empty host integument.

Hosts: in Britain *Orthosia cerasi* (Lep.: Noctuidae) on *Quercus* (UMO). In Europe single records from this host, *Peridea anceps* (Lep.:Notodontidae), *Orthosia cruda, Cucullia verbasci* (both Lep.: Noctuidae), *Malacosoma neustria* and the questionably British *Dendrolimus pini* (both Lep.:Lasiocampidae).

British distribution: S.England and (southern) Midlands.

Habitat: woodland margins (Herting, 1960).

Flight period: early April to early July (at least 50 records).

T. ursina. Pupates in the empty host integument.

Host: unknown.

British distribution: England, Wales (Glamorgan and Flint), S.Scotland (Edinburgh* and Kirkcudbright*) and N.Scotland (Gartmorn Dam in Perths*).

Habitat: woodland margins (Herting, 1960; Chandler, pers. comm.)

Flight period: late March to early May (at least 50 records).

Genus **Nowickia**

N. ferox

Hosts: in Britain *Apamea monoglypha* (Lepidoptera: Noctuidae) (NHM). In Europe single records from this host and (questionably) *Xylena exsoleta* (Lep.: Noctuidae).

British distribution: S.England, Midlands and N.England (Lancs*).

Habitat: heathland and woodland margins (Herting, 1960).

Flight period: early July to late September (at least 50 records).

Genus **Peleteria**

P. rubescens

Hosts: in Europe: *Euxoa* spp (4) and *Agrotis* spp (8) (both Lepidoptera: Noctuidae); *Euproctis chrysorrhoea* (1) (Lep.: Lymantriidae); *Papilio machaon* (1) (Lep.: Papilionidae) and (questionably) *Malacosoma castrensis* (1) (Lep.: Lasiocampidae).

British distribution: S.England (Burnham Beeches in Bucks 20.ix.1931 NHM; Devil's Ditch in Cambs*).

Habitat: in Europe found in grassland (Herting, 1960).

Flight period: in Europe end of May to September (Herting, 1960).

Genus **Germaria**

A single record of another species in this genus reared from the non-British *Bradyrrhoa gilveolella* (Treitschke) (Lepidoptera: Pyralidae).

G. angustata. The most recent record is 1947.

Hosts: unknown.

British distribution: S.England (Horsey, Waxham, Blakeney Point, Holme and Winterton Sandhills in Norfolk; Faversham* and Martham in Kent; unknown locality in Essex).

Habitat: localities are sandy coastal areas (Falk, in press). In Europe also found in coastal areas (Herting, 1960).

Flight period: late June to early August (18 records). In Europe May to August (Herting, 1960).

G. ruficeps. The most recent record is 1950.
Hosts: unknown.
British distribution: S.England (Breamore in Hants vii.1950 BENHS; Faversham 31.vii.1904 UMO and Dover* in Kent; Yarmouth in Norfolk*).
Flight period: in Europe July and August (Herting, 1960).

Tribe Nemoraeini

Genus **Nemoraea**

N. pellucida
Hosts: in Europe a range of Lepidoptera larvae. In Britain: *Orthosia cerasi* (Noctuidae) (Wainwright, 1928) and either *Hyloicus pinastri* (Sphingidae) or *Dypterygia scabriuscula* (Noctuidae) (NHM).
British distribution: S.England (unknown locality in S.Devon; Yellowham Wood in Dorset; Guestling in Sussex; New Forest, Breamore, Micheldever and Liphook in Hants).
Habitat: localities indicate woodland margins (Falk, in press). In Europe usually found in woodland (Herting, 1960).
Flight period: early June to early September (15 records). In Europe 2 generations − May/June and August (Tschorsnig, pers. comm.).

Tribe Linnaemyiini

Genus **Linnaemya**

L. comta. First-instar larvae can survive for an average of 7.5 days on some plant surfaces (Clement *et al.*, 1986). This enables them to attack hosts which are in the soil during the day, only emerging to feed at night. They are active, able to move up to 8cm. Pupation occurs outside the host in the ground.
Hosts: in Europe single records from *Agrotis segetum, A.ipsilon, Agrotis* sp. and *Euxoa tritici* (all Lepidoptera: Noctuidae: Noctuinae).
British distribution: S.England (south of a Bristol-London line) and N.Scotland (N.Sutor in Ross; Glenmore in Inverness; Findhorn in Moray*).
Habitat: localities are heathland, downland, wastes and woods (Falk, in press).
Flight period: early June to early August (20 records). In Europe June to September, possibly 2 generations per year (Herting, 1960).

L. vulpina
Hosts: in Britain *Lycophotia porphyrea* (Lepidoptera: Noctuidae) (Ford, 1973; Ford, 1976; Ford & Shaw, 1991). In Europe 3 further records from this host and single records from *Chilodes maritimus, Archanara geminipuncta* and *Blepharita satura* (all Noctuidae).
British distribution: England, Wales, S.Scotland (Wigtown and Renfrew), N.Scotland and Ireland (Kerry*, Cork, Mayo, Wicklow, Donegal*, Derry * and Waterford*).

Habitat: wet *Erica* heathland (Day, 1948; Herting, 1960). Hosts feed on heather and reeds.

Flight period: early July to early September (at least 50 records).

L. tessellans. First-instar larvae are attached posteriorly to the leaf by the egg case (see fig.307).

Hosts: in France a questionable record from *Xestia c-nigrum* (Lepidoptera: Noctuidae).

British distribution: S.England.

Habitat: usually woodland or scrub (Herting, 1960).

Flight period: 2 periods – late May/June and August/early September (at least 50 records).

L. rossica

Hosts: in Britain *Xestia agathina* (Lepidoptera: Noctuidae) (Ford & Shaw, 1991). Hosts in Emden (1954) assumed to be erroneous.

British distribution: S.England (Woolwich Wood in Kent), N.England and N.Scotland.

Habitat: in Europe usually found in cool montane areas with pine forests (Tschorsnig, pers. comm.).

Flight period: July and August (at least 50 records).

Genus **Chrysocosmius**

C. auratus

Hosts: in Europe: *Horisme tersata* (1) and the non-British *Melanthia alaudaria* (Freyer) (3) (both Lepidoptera: Geometridae).

British distribution: S.England (South Rodborough in Gloucs 17.vii.1943 BCM). Commoner in southern Europe but range extends to central Sweden and Finland (Tschorsnig, pers. comm.).

Flight period: July and August (19 European records).

Genus **Lydina**

L. aenea

Hosts: in Europe single unconfirmed records from *Zeiraphera diniana* (Lepidoptera: Tortricidae), *Acrobasis repandana* (Lep.: Pyralidae), *Eupithecia* sp. (Lep.: Geometridae) and *Panolis flammea* (Lep.: Noctuidae).

British distribution: England, Wales, N.Scotland and Ireland (Kerry, Cork, Louth and Clare).

Flight period: 2 periods – May/early June and late July/August (at least 50 records).

Genus **Lypha**

L. dubia. First-instar larvae are mobile but periodically stop and make pendulum-like host-searching motions with their anterior body. When attacking *Rhyaciona buoliana* the eggs are laid on the host web. Leaves the host to overwinter as an adult within the puparium in the soil. Biology and immature stages in Cheng (1969). Ecology in Schröder (1969) and Cheng (1970).

Hosts: tree-feeding Lepidoptera larvae. In Britain: Chrysomelidae (Coleoptera) – *Chrysolina* sp. (Cheng, 1969 – probably erroneous). Syrphidae (Diptera) – *Merodon ?equestris* (Collin, 1945 – probably erroneous). Geometridae (Lepidoptera) – *Agriopis marginaria* (Carr, 1935), *Erannis defoliaria* (Ford, 1976), *Operophtera brumata* (6 – UMO inc. 3 on *Quercus* and 1 on *Corylus avellana*). Tortricidae – *?Archips podana* on *Quercus* (UMO), *Tortricodes alternella* on *Quercus* (6 – UMO), *Zeiraphera isertana* on *Quercus* (UMO). Also recorded from an unidentified Tortricidae on *Pteridium* (Ford & Shaw, 1991).

British distribution: England, Wales, S.Scotland (Stirling*), N.Scotland (Inverness and Moray) and Ireland (unknown locality).

Habitat: collection localities and hosts indicate woodland.

Flight period: early April to early June (at least 50 records). 1 generation per year.

L. ruficauda

Hosts: chiefly *Hydriomena* larvae (Lepidoptera: Geometridae). In Europe: *H.impluviata* (6) and *H.ruberata* (2), also questionably *Tritophia tritophus* (Lep.: Notodontidae). In Britain recorded from *H.impluviata* (Geometridae) (UMO) and *Acronicta alni* (Lep.: Noctuidae) (UMM). Also recorded from an unidentified host on *Narcissus* (NHM).

British distribution: S.England (Shipload Bay in Devon; Farley Downs in Hants; Hickling in Norfolk), N.England and N.Scotland.

Habitat: moorland birches (Emden, 1954). In Europe usually found in cool montane areas with pine forests (Tschorsnig, pers. comm.).

Flight period: June and July (7 records). In Europe mid-June to mid-August (Herting, 1960).

Tribe Ernestiini

Genus Ernestia

E. laevigata. Overwinters as a puparium in the soil.

Hosts: in Britain: Noctuidae – *Abrostola triplasia* (NHM), *Cosmia trapezina* on *Quercus* (2 – UMO), *Orthosia cruda* on *Quercus* (2 – UMO), *O.cerasi* on *Quercus* (UMO). Thyatiridae – *Achlya flavicornis* (NHM) (no locality data). In Europe further records from *Orthosia* spp (3) and *C.trapezina* (2). All hosts excluding *A.triplasia* are tree-feeders.

British distribution: S.England, Midlands, Wales (Glamorgan and Merioneth) and N.Scotland (Moray and Aberdeen).

Habitat: woodland (Herting, 1960).

Flight period: late April to early June (at least 50 records).

E. puparum

Hosts: unknown. Records in Emden (1954) probably erroneous (Herting, 1960).

British distribution: S.England (Hants, Berks, Surrey and Herts*) and Midlands (Notts* and Lincs*).

Habitat: localities indicate broadleaved woodland (Falk, in press). In Europe found in woodland margins (Herting, 1960).

Flight period: late March and April (23 records).

E. rudis. The first-instar larva remains attached to the leaf by its egg case (fig.307) for up to 11 days. It responds to vibrations with a pendulum-like movement of its anterior body until a host is contacted. Solitary, leaving the host to pupate in the soil.

Hosts: in Europe a regular parasitoid of the pine-feeding *Panolis flammea* (Noctuidae), also recorded from *Orthosia* spp (3), *Melanchra persicaria* (1) and *Xestia c-nigrum* (1) (all Noctuidae). In Britain *Orthosia incerta* (UMM).

British distribution: England, Wales (Merioneth and Flint), N.Scotland (Inverness, Perth* and Ross*) and Ireland (Waterford*).

Habitat: deciduous and coniferous woodland (Herting, 1960).

Flight period: May and June (at least 50 records).

E vagans

Hosts: in Britain: *Achlya flavicornis* (2) and *Polyploca ridens* (2) (both Lepidoptera: Thyatiridae). In Europe a further 4 records from the above 2 species.

British distribution: England, Wales (Merioneth and Cardigan), S.Scotland (Stirling*) and N. Scotland (Inverness and Moray).

Habitat: usually deciduous woodland (Herting, 1960).

Flight period: May and June (at least 50 records).

Genus **Appendicia**

A. truncata

Hosts: in Europe one record from *Cerapteryx graminis* (Lepidoptera: Noctuidae).

British distribution: S.England (Somerset, Gloucester, Wilts, Sussex and London), Midlands, Wales, N.England and N.Scotland.

Habitat: mainly northern moorlands (Emden, 1954). An alpine-boreal species (Herting, 1960).

Flight period: May and June (at least 50 records).

Genus **Fausta**

F. nemorum

Hosts: in Britain *Orthosia cruda* (Lepidoptera; Noctuidae) (Audcent, 1942). In Japan a probably erroneous record from a Cimbicidae sawfly (Hymenoptera: Symphyta).

British distribution: S.England, Midlands (Hereford) and Wales (Pembroke).

Habitat: woodland (Herting, 1960). [Downland in Emden (1954).]

Flight period: July and August (at least 50 records).

Genus **Eurithia**

E. anthophila. The first-instar larva behaves like *Ernestia rudis* (see above) but it is more active, being able to rapidly move several cm if disturbed. Overwinters as a puparium in the ground.

Hosts: Lepidoptera larvae. In Britain: Arctiidae – *Spilosoma lubricipeda* or *lutea* (NHM). Noctuidae – ?*Nonagria typhae* (LM), *Lacanobia oleracea* (Ford, 1976). Also from a puparium found among pine litter (NHM). In Europe a further 2 records from *S.lubricipeda* (1 possibly refers to *lutea*), also from *Ptilodon capucina* (1) (Notodontidae) and *Melanchra persicariae* (1) (Noctuidae).

British distribution: England, Wales, S.Scotland (Dunbarton), N.Scotland (Inverness) and Ireland (Kerry, Cork, Waterford, Down*, Fermanagh* and Mayo).

Habitat: woodland (Day, 1948; Chandler, pers. comm.) and meadows (Herting, 1960).

Flight period: late July to early September (at least 50 records). 1 generation per year.

E. caesia

Hosts: in Europe: *Hadena* spp (6) and *Noctua pronuba* (1) (both Lepidoptera: Noctuidae). In Britain 1 questionable record from *Hadena* sp. on *Silene* or *Melandrium* ("Campion").

British distribution: S.England (Knighton woods in Dorset; Portsdown and Breamore in Hants; Chipperfield in Herts; Wrotham, Folkestone and St. Margaret in Kent).

Flight period: early June to early August (5 records). In Europe June and July (Herting, 1960).

E. intermedia

Hosts: unknown.

British distribution: S.England, Wales (Glamorgan) and N.England (Lancs).

Flight period: late April to early June (22 records including 3 possibly in July/August). In Europe end of April to the beginning of June (Herting, 1960).

E. connivens

Hosts: no British records. Single Palaearctic records from *Lymantria dispar* (Lepidoptera: Lymantriidae), *Euplexia lucipara* (Lep.: Noctuidae) and probably erroneously the non-British *Choristoneura murinana* (Hübner) (Lep.: Tortricidae).

British distribution: S.England (Cornwall, Devon, Somerset* and Gloucs*), Midlands (Cheshire), Wales (Glamorgan*), N. England (Lancs) and Ireland (Kerry, Galway, Wicklow, Waterford, Down*, Fermanagh* and Mayo).

Flight period: July and August (8 records). In Europe mid-July to the beginning of September (Herting, 1960).

E. consobrina.

The first-instar larva is attached to the leaf in the same manner as *Ernestia rudis*, surviving for an average of 30 hours.

Hosts: chiefly Hadeninae larvae (Lepidoptera: Noctuidae). In Britain: Hadeninae – *Amphipyra tragopoginis* (Ford, 1976), *Ceramica pisi* (Ford, 1973), *Lacanobia oleracea* (Ford, 1976; Ford & Shaw, 1991). Cuculliinae (Noctuidae) – *Xanthia gilvago* (Ford & Shaw, 1991). In Siberia also recorded from *Lymantria dispar* (Lep.: Lymantriidae) (Sabrosky & Reardon, 1976).

British distribution: S.England, Midlands (Warwick and Staffs), Wales (Glamorgan), N.England, N.Scotland (Ross* and Aberdeen*) and Ireland (Kerry, Cork and Waterford).

Flight period: late July to early September (80% of records) with some in May and June (at least 50 records). 2 generations per year.

E. vivida

Hosts: in Europe single records from *Lithophane lamda* and the non-British *Orthosia opima* (Hübner) (both Lepidoptera: Noctuidae).

British distribution: S.England and N.Scotland (Fetlar in Shetland).

Habitat: in Europe usually found in cool montane areas with pine forest (Tschorsnig, pers. comm.). [Downland (Emden, 1954).]

Flight period: late June to early September (16 records).

Genus **Hyalurgus**

H. lucidus. The first-instar larva is not heavily sclerotised, possessing ambulatory belts of spines.

Hosts: Nematinae sawfly larvae (Hymenoptera: Symphyta: Tenthredinidae), attacking species on shrubs and trees (Pschorn-Walcher, 1969). No British records.

British distribution: Midlands (Wyre Forest in Worcs; Woolhope, Shobdon Marshes, Nash and Ashperton Park in Hereford), Wales (Rhyader in Radnor) and N.Scotland (Logie in Moray).

Habitat: woodland (Herting, 1960; Falk, in press).

Flight period: late June and July (11 records). In Europe July and August (Herting, 1960).

Genus **Gymnocheta**

G. viridis. The first-instar larva is less heavily sclerotised than *Ernestia rudis*, indicating that it is probably more mobile.

Hosts: in Europe single records from *Photedes minima*, *P.pygmina* and *Mesapamea secalis* (all Lepidoptera: Noctuidae). These are stem borers in grasses or (*P.pygmina*) sedges. Also a questionable record from *Scotopteryx chenopodiata* (Lep.: Geometridae). A British record from *Calliteara pudibunda* (Lep.: Lymantriidae) on *Quercus* sp. (Hammond & Smith, 1953) is probably erroneous given the host phenology (see South, 1961).

British distribution: England, Wales, N.Scotland and Ireland (Cork and Dublin).

Habitat: woodland, the males often on sunlit tree trunks (Herting, 1960; Chandler, pers. comm.).

Flight period: early April to early June (at least 50 records).

Genus **Zophomyia**

Z. temula. The first-instar larva adopts the host-searching posture shown in fig.307, but is also able to move short distances using a looping motion.

Hosts: unknown.

British distribution: S.England, Midlands (Worcs), Wales and N.England (Westmorland*).

Habitat: localities are woodland, calcareous grassland and coastal dunes (Falk, in press).

Flight period: late May to early July (at least 50 records).

Genus **Cleonice**

C. callida. Larva usually leaves the host (often after it has pupated) to overwinter as a puparium in the soil.

Hosts: Chrysomelidae larvae (Coleoptera). In Europe: *Chrysomela populi* (7), *C.tremula* (1) and the non-British *C.saliceti* (Ws.) (1).

British distribution: S.England (Bricket Wood in Herts 22.vi.1947 NHM).

Habitat: in Europe found on *Populus tremula* (Herting, 1960).

Flight period: in Europe May and June (Herting, 1960).

Genus **Loewia**

L. foeda. The first-instar larva is very heavily sclerotised. Pupates in empty host integument. See also Wood & Wheeler (1972).

Hosts: in Europe recorded from the centipede *Lithobius* sp. (Chilopoda) (Thompson, 1915).

British distribution: S.England, Midlands (Hereford* and Notts*) and N.England (York*).

Flight period: July and August (28 records).

L. submetallica

Hosts: unknown.

British distribution: S.England, Midlands (Hereford) and N.England (Lancs).

Flight period: late June to late August (19 records).

L. phaeoptera

Hosts: unknown.

British distribution: S.England, Midlands (Worcs and Hereford) and Wales (Monmouth*).

Flight period: July and August (45 records).

Genus **Eloceria**

E. delecta. The first-instar larva is heavily sclerotised. Overwinters in host, pupating in the empty host integument.

Hosts: in Britain the centipede *Lithobius* (Chilopoda) (NHM). In Europe a further 2 records from this host.

British distribution: S.England, Wales (Glamorgan) and N.England (York*).

Flight period: early June to early August (40 records).

Tribe **Pelatachinini**

Genus **Pelatachina**

P. tibialis. The first-instar larva is not heavily sclerotised and does not make host-searching movements, indicating that the egg is probably laid on the host. Usually solitary, leaving the host to pupate in the soil and usually overwintering as a puparium.

Hosts: chiefly Nymphalidae larvae (Lepidoptera), the majority of records from *Inachis io* and *Aglais urticae*. In Europe also a few records from Noctuidae (Lep.). In Britain: Nymphalidae – *Aglais urticae* (5), *Eurodryas aurinia* (1 – UMM and LM), *Inachis io* (Ford, 1976). Also an unidentified Geometridae (Lep.) (UMO).

British distribution: England, Wales (Glamorgan) and N.Scotland.

Flight period: May and June (at least 50 records). In Europe occasional specimens in July and August indicating a partial second generation (Tschorsnig, pers. comm.).

Tribe Macquartiini

Genus **Macquartia**

All host records are from Chrysomelidae larvae (Coleoptera). The record of *M.tessellum* from Britain (Richards & Waloff, 1959) is probably erroneous, being far outside its range (Herting, 1984).

M. dispar
Hosts: in Europe questionable single records from *Chrysolina sanguinolenta* and *Timarcha goettingensis* (both Coleoptera: Chrysomelidae).

British distribution: S.England, Midlands (Worcs) and Ireland (Down*, Antrim*, Wicklow, Clare and Kerry).

Flight period: 2 periods – late April to early June (90%) and late August to late October (10%) (at least 50 records).

M. grisea
Hosts: in Britain *Chrysolina oricalcia* (Coleoptera: Chrysomelidae) on *Anthriscus cerefolium* (NMW). In Europe single records from *C.sanguinolenta* and *C.fastuosa*.

British distribution: S.England, Midlands, Wales (Monmouth) and N.England (Lancs).

Flight period: late April to early October (at least 50). At least 3 generations per year.

M. nudigena
Hosts: unknown.

British distribution: S.England (Gloucs* and Cambridge*), Midlands, Wales (Merioneth*), N.England (Westmorland) and N.Scotland.

Flight period: May and June (21 records).

M. praefica
Hosts: in Europe a record from *Chrysolina varians* (Coleoptera: Chrysomelidae).

British distribution: England, Wales (Pembroke and Glamorgan) and Ireland (Mayo).

Flight period: early June to late August, peaking late June and early August (at least 50 records).

M. pubiceps
Hosts: a record from the Indian *Chrysolina aurata* (Suffr.) (Coleoptera: Chrysomelidae) (Emden, 1950).

British distribution: England, Wales (Glamorgan), S.Scotland (Wigtown and Dumfries) and N.Scotland (Inverness and Moray).

Flight period: early May to late August (at least 50 records). Possibly 3 generations.

M. tenebricosa. The egg is usually laid on the host but the larva is capable of searching on the host plant. Solitary, overwintering as a second-instar larva in the host. Host survives to construct its pupation cell in the soil and the tachinid pupates in the empty host integument.

Hosts: in Britain: *Chrysolina graminis* on *Tanacetum vulgare* (Drummond, 1952) and *C. varians* (5) (Coleoptera: Chrysomelidae). In Europe also other *Chrysolina* spp (4).

British distribution: England, Wales and N.Scotland (Inverness, Sutherland and Clyde Isles).

Flight period: late May to late September (at least 50 records). In Europe 2 or 3 generations per year (Herting, 1960).

M. viridana

Hosts: in Europe a record from the non-British *Colaphellus sophiae* (Schall.) (Coleoptera: Chrysomelidae).

British distribution: S.England, Midlands (Hereford and Derby) and Wales (Glamorgan and Camarthen*).

Flight period: early April to early June (33 records).

Genus **Anthomyiopsis**

The southern European *A. plagioderae* (with which the British species has been confused) lays eggs on the larvae of its chrysomelid host. Overwinters as a first-instar larva in the host and emerges from the adult the following autumn, pupating on the host food-plant.

A. nigrisquamata

Hosts: in Europe single records from *Phyllodecta vitellinae* and (questionably) from the non-British *Colaspidema atrum* Ol. (both Coleoptera: Chrysomelidae).

British distribution: S.England (Barnridge Copse*, Bentley Woods* and Winterslow all in Wilts) and N.Scotland (Loch Vennachar in Perth; several localities in Spey Valley).

Habitat: localities indicate aspen copses (Falk, in press). Numerous on aspen (Emden, 1954).

Flight period: late June/early July (20 records).

Tribe **Triarthriini**

Genus **Triarthria**

T. setipennis. The first-instar larva is covered in sclerotised plates and spines, actively searching for the host. Leaves the host to pupate nearby in crevices, overwintering as a puparium.

Hosts: the earwig *Forficula auricularia* (Dermaptera), see Phillips (1983). In Europe also a record from the non-British *Chelidura albipennis* (Dermaptera). Scattered records from Lepidoptera (usually stem-borers or root-feeders) are probably the result of erroneous associations (Herting, 1960).

British distribution: England, Wales, N.Scotland and Ireland (Down*).

Flight period: late April to early August, peaking June (at least 50 records).

Tribe Neaerini

Genus Neaera

N. laticornis

Hosts: in Britain *Eucosma hohenwartiana* (Lepidoptera: Tortricidae) on *Centaurea scabiosa* (Parmenter, 1953). In the Soviet Union a record from *Pexicopia malvella* (Lep.: Gelechiidae).

British distribution: S.England.

Habitat: in southern England found in chalk areas (Herting, 1960). The British host is a concealed feeder associated with grassland (Bradley *et al.*, 1979).

Flight period: July and August (45 records).

Genus Phytomyptera

The European species of this genus have recently been revised by Andersen (1988). The occasional presence of rows of empty eggs in the distal part of the uterus indicates that the genus may be larviparous. As many hosts are concealed it is likely that the first-instar larva contacts the host rather than the egg being laid on it.

P. cingulata

Hosts: chiefly microlepidoptera larvae in rotting wood, fungi or lichens. In Britain: Gelechiidae – *Teleiodes sequax* (Ford & Shaw, 1991). Oecophoridae – *Esperia sulphurella* (Audcent, 1932; Ford & Shaw, 1991 – on *Prunus*), an unidentified species on *Quercus* (Ford & Shaw, 1991). Tineidae – *Nemapogon cloacella* from the bracket fungus *Phellinus pini* (Brot. per Fr.) on *Pinus* (Ford & Shaw, 1991) and from the bracket fungus *Polyporus squamosus* Huds. ex Fr. (NHM), *N.variatella* from a bracket fungus on *Quercus* (Ford & Shaw, 1991), *Nemaxera betulinella* (NHM), *N.granella* (Emden, 1954). Tortricidae – *Acleris variegana* on *Berberis* (Ford & Shaw, 1991) and an unidentified species on *Teucrium* (Ford & Shaw, 1991). Also records from unidentified hosts on *Malus* (Ford & Shaw, 1991) and *Betula* (UMO), and from either *Schiffermuelleria similella* (Oecophoridae) or *Nemopogon cloacella* (Tineidae) on rotting *Pinus* (Ford & Shaw, 1991). In Europe also a record from *Dahlica triquetrella* (Psychidae).

British distribution: England, Wales (Montgomery and an unknown locality in North Wales) and N.Scotland.

Flight period: late May to early September (at least 50 records).

P. nigrina. Overwinters as a second-instar larva in the host, pupating in the empty host pupa.

Hosts: stem-boring and concealed microlepidoptera larvae. In Britain: Pterophoridae – *Adaina microdactyla* (Ford & Shaw, 1991; BCM – in stem of *Eupatorium cannabinum*). Tortricidae – *Archips rosana* (Ford & Shaw, 1991), *Epinotia immundana* (3 – inc. 1 on *Alnus*). Also recorded from an unidentified host in the stem of *Agrimonia* sp. (LM) and *Eupatorium cannabinum* (2).

British distribution: S.England, Midlands (Hereford and Warwick*), Wales (Merioneth*), N.Scotland and Ireland (Kerry).

Flight period: late May to late August (44 records). More than 1 generation per year.

Genus **Graphogaster**

G. brunnescens
Hosts: in Europe single records from *Leucoptera laburnella* (Lepidoptera: Lyonetiidae), *Teleiodea notatella* (Lep.: Gelechiidae), *Acleris ferrugana* and *Retinia resinella* (both Lep.: Tortricidae).

British distribution: S.England (Wilts, Gloucs and Kent) and N.Scotland.

Flight period: late June to late August (14 records). In Europe end of June to beginning of September (Herting, 1960).

Tribe **Siphonini**

This group has recently been studied in detail (Andersen, 1983; O'Hara, 1988a, 1988b and 1989). Hosts are a variety of Lepidoptera larvae except for some *Siphona* which attack Tipulidae larvae (Diptera). All species lay incubated eggs, hatching at or possibly just prior to deposition. The structure of the ovipositor suggests that, except in *Actia, Siphona* and *Aphantorhaphopsis*, the female oviposits directly onto the host (Andersen, 1983). Many *Actia* attack stem-boring or otherwise concealed hosts, presumably laying their eggs at the gallery entrance.

Genus **Goniocera**

G. versicolor. Overwinters as a puparium.

Hosts: in Britain *Malacosoma neustria* (Lepidoptera: Lasiocampidae) (5 – inc. 1 on *Crataegus*). In Europe a record from *M.castrensis* and, in Siberia, from *Aporia crataegi* (Lep.: Pieridae) which is extinct in Britain.

British distribution: S.England (Yealm Mouth in Devon; Cheltenham in Gloucs; Langley in Bucks; Colchester in Essex; Sevenoaks in Kent) and Wales (near Cardiff in Glamorgan).

Flight period: May (4 – inc. 3 eclosion dates). In Europe May (Herting, 1960).

Genus **Entomophaga**

E. exoleta
Hosts: in Europe a questionable record from an unidentified Geometridae (Lepidoptera).

British distribution: S.England (Coombe Wood near Wool in Dorset 8.v.1949 NHM).

E. nigrohalterata. Overwinters as a puparium in the ground (Cheng, 1967).

Hosts: in Britain the web-forming microlepidoptera *Ypsolopha parenthesella*, *Y.ustella* and *Y.alpella* (Yponomeutidae) (UMO – all on *Quercus robur*). No European records.

British distribution: S.England, Midlands (Hereford and Worcs) and N.England (Westmorland).

Habitat: in Europe usually found in deciduous woodland (Andersen, pers. comm.)

Flight period: May and early June (26 records).

Genus **Ceromya**

C. bicolor. Gregarious in larger hosts; 28 adults have been reared from a single individual. Overwinters as a puparium.

Hosts: chiefly Lasiocampidae larvae (Lepidoptera). In Britain: Lasiocampidae – *Lasiocampa quercus* (12 – inc. 1 on *Salix ?atrocinerea*), *L.trifolii* (2). Geometridae – *Eupithecia* sp. (Ford, 1973). In Europe: other Lasiocampidae, *Phragmatobia fuliginosa* (Arctiidae) and (questionably) *Hyloicus pinastri* (Sphingidae).

British distribution: England, Wales and N.Scotland.

Flight period: early May to early August (at least 50 records).

C. monstrosicornis
Hosts: unknown.

British distribution: S.England (Shotover in Oxford 6.v.1916 UMO; Felden in Herts 13.vi.1894 NHM; Cambridge 26.v.1905 (Wainwright, 1928); Monks Soham in Suffolk 17.v.1940 NHM) and an unknown locality in England (NHM).

C. silacea
Hosts: in Europe a record from *Protodeltote pygarga* (Lepidoptera: Noctuidae) – an exposed grass-feeder.

British distribution: S.England (Brockenhurst Wood in Hants vii.1990 (Chandler, pers. comm.); Chippenham Fen in Cambs 1932-1974 NHM and UMO).

Flight period: late June and July (8 records).

Genus **Actia**

A. crassicornis
Hosts: concealed microlepidoptera larvae (Oecophoridae and Tortricidae), the majority of confirmed records from *Depressaria*. Easily confused with *pilipennis*, to which the Audcent (1932) pterophorid and tortricid host records are thought likely to refer (Herting, 1960). In Britain: Oecophoridae – *Agonopterix conterminella* (Ford & Shaw, 1991) and *A.heracliana* (UMO). Also an unidentified host on *Angelica* (Richards, 1935).

British distribution: S.England, Midlands (Salop), Wales (Merioneth* and Flint), N.England, N.Scotland and Ireland (Donegal*, Kildare and Wicklow).

Flight period: late May to early August (at least 50 records). More than 1 generation per year.

A. infantula.
Hosts: in Britain the saprophagous *Monopis laevigella* (Lepidoptera: Tineidae) (2 – inc. Woodroffe, 1953). Also 2 records from unknown hosts in bird nests. No European records.

British distribution: S.England and Midlands (Hereford and Worcs).

Flight period: late June to late August (46 records).

A. lamia
Hosts: *Epiblema* larvae (Lep.: Tortricidae), stem- and root-boring and concealed feeders on shrubs and herbs. In Britain recorded from *E.foenella* in the stem of *Cirsium palustre* (Parmenter, 1953 – but see Emmet, 1988) and *E. scutulana* (Audcent, 1932; Uffen, 1961 – in stem of *Cirsium palustre*). Also recorded from an unidentified Tortricidae in a thistle stem (UMO). References to *Lasiocampa*

quercus and *Ourapteryx sambucaria* as hosts are erroneous (from misreadings of Wainwright, 1928: 208).

British distribution: S.England, (southern) Midlands, Wales (Monmouth and Glamorgan) and N.Scotland (Grantown in Moray).

Flight period: early May to late July (32 records). More than 1 generation per year.

A. maksymovi

Hosts: in Europe concealed Tortricidae larvae on *Larix* and *Abies*.

British distribution: S.England (Alice Holt Park in Hants eclosed 7.v.1965 NHM).

A. nudibasis. Pupates within the entrance of the host gallery.

Hosts: *Retinia resinella* and *Rhyacionia buoliana* on *Pinus* (both Lepidoptera: Tortricidae), within developing buds and galls respectively. Also a few records from other microlepidoptera on *Pinus*. The first generation of *A. nudibasis* attacks *R. resinella* and the second *R. buoliana* (Herting, 1960). In Britain one record from *R. buoliana* (Wainwright, 1932).

British distribution: S.England (Aldershot in Hants 11.vi.1920 NHM; Weybridge 10.vii.1911 UMO and Ockham 21.v.1966 NHM in Surrey; unspecified locality in Suffolk*).

A. pilipennis. Pupates within the spun leaves of the host.

Hosts: a wide range of stem-boring and concealed microlepidoptera. In Britain: Choreutidae – *Anthophila fabriciana* (NHM on nettle; 3 in Ford & Shaw, 1991). Gelechiidae – *Hypatima rhomboidella* (Ford & Shaw, 1991). Oecophoridae – *Agonopterix ulicetella* on *Ulex minor* (Ford & Shaw, 1991). Pterophoridae – *Pterophorus pentadactyla* (UMM). Tortricidae – *Acleris aspersana* (Emden, 1954), *Acleris rufana* (Ford & Shaw, 1991), *Acleris logiana* (UMO), *Aphelia viburnana* on *Pinus contorta* or *Abies procera* (Winter, 1974), *?Archips podana* on *Laburnum* (UMM), *Croesia bergmanniana* (Emden, 1954), *Cydia gemmiferana* (NHM), *Ditula angustiorana* on *Malus* (Hey, 1935), *Endothenia pullana* (NHM), *Lozotaenia forsterana* (Ford & Shaw, 1991), *Pandemis corylana* (Matthey, 1967) on *Quercus* (Cheng, 1967), *Pandemis cerasana* on *Quercus robur* (UMO), *Ptycholoma lecheana* on *Quercus robur* (UMO),. *Tortrix viridana* (NHM; 3 UMO on *Quercus robur*). Also from unidentified hosts on *Hedera* and *Spiraea salicifolia*.

British distribution: England, Wales, S.Scotland (Selkirk*), N.Scotland and Ireland (Kerry*).

Flight period: end of May to late August, peaking late June and early August (at least 50 records).

Genus **Peribaea**

P. fissicornis. Gregarious

Hosts: Geometridae larvae (Lepidoptera) on trees and shrubs. In Britain: Geometridae – *Agriopis aurantiaria* on *Corylus* (UMO) and *Ourapteryx sambucaria* (NHM). Lasiocampidae – *Poecilocampa populi* on *Betula* (Hammond & Smith, 1957).

British distribution: S.England, Midlands (Hereford and Warwick) and N.Scotland (Loch Coire in Inverness; Grantown in Moray).

Habitat: localities indicate woodland (Falk, in press).

Flight period: early May to early August (33 records). More than 1 generation per year.

Genus **Ceranthia**

C. abdominalis

Hosts: in Europe: *Cyclophora* spp (7) and *Thera britannica* (1), both Geometridae (Lepidoptera) feeding on leaves of deciduous trees. In Britain *C.annulata* (Audcent, 1932).

British distribution: S.England (Hants and Hunts), Midlands (Hereford), Wales (Merioneth*) and N.Scotland (Spey valley in Moray/Inverness; Cambus O'May in Aberdeen).

Flight period: July and August (15 records).

C. lichtwardtiana

Hosts: in Britain *Eupithecia* sp. (Lepidoptera: Geometridae) on *Betula* (Ford & Shaw, 1991). In Europe one record from *Acasis viretata* (Lep.: Geometridae).

British distribution: S. England (Oxon), Midlands (Salop), Wales (Flint) and N.Scotland.

Flight period: June and July (16 records).

Genus **Aphantorhaphopsis**

A. verralli

Hosts: unknown.

British distribution: Midlands (Whixall Moss in Salop*) and N.Scotland (near Grantown in Moray 4.viii.1935 UMO and 13.vii.1938 NHM; The Mound in Sutherland 26.vii.1914 UMO).

Genus **Siphona**

This extremely difficult genus has recently been revised by Andersen (1982). The number of species recognised in Britain has increased from 3 in Emden (1954) to 10. Literature records should therefore be treated with caution. Certain species may actually be commoner and more widely distributed than is indicated here.

S. boreata

Hosts: unknown.

British distribution: S.England (Silwood Park in Berks), N. Scotland (Lochloyal in Sutherland; Spey Valley in Inverness/Moray) and Ireland (Galway).

Flight period: late May to early July (4 plus a series of 19 specimens taken in Malaise traps operating all season at Silwood Park during 1989).

S. collini

Hosts: in Britain an unidentified Noctuidae (Lepidoptera) (UMM). In Europe *Cerapteryx graminis* and *Euxoa tritici* (both Lep.: Noctuidae), nocturnal feeders on grass and herbs respectively.

British distribution: S.England, Midlands (Warwick), Wales (Glamorgan), N.England (Lancs) and N. Scotland.

Flight period: early May to early August (38 records).

S. cristata. Gregarious, overwintering as a puparium in the ground.

Hosts: mainly nocturnally-feeding Noctuidae larvae (Lepidoptera). In Britain: Noctuidae – *Agrotis ipsilon* (NHM), *Caradrina morpheus* (Hammond & Smith, 1955), *Lacanobia oleracea* on *Calendula* (Hammond & Smith, 1955), *Mamestra brassicae* (9), *Mythimna litoralis* (2 – UMO), *Phlogophora meticulosa* (BCM). Also from an unidentified Noctuidae on *Clematis montana* (Ford & Shaw 1991 – determined by S.Andersen). Geometridae – *Erannis defoliaria* (NHM). Sphingidae – *Smerinthus ocellata* (Hammond & Smith, 1955).

British distribution: England, Wales (Glamorgan), S.Scotland (Edinburgh*) and N.Scotland.

Flight period: late June to early August (at least 50 records).

S. geniculata. Often gregarious, overwintering in the host larva (Rennie & Sutherland, 1920). Life history in Alma (1976).

Hosts: Tipulidae larvae (Diptera). In Britain *Tipula paludosa* (4). Records from Lepidoptera (*Ceramica pisi* (Ford, 1973) and *Pieris brassicae* (BCM)) are questionable.

British distribution: England, Wales, S.Scotland (Dumfries), N.Scotland and Ireland (Monaghan* and all coastal counties from Kerry east to Dublin excluding Wexford).

Flight period: late May to late September (at least 50 records). 2 generations per year observed in Scotland (Rennie & Sutherland, 1920) and 3 in S.England (author, pers. obs.) and Denmark (Andersen, 1982).

S. ingerae

Hosts: unknown.

British distribution: S.England (Somerset and Oxon) and Midland (Worcs).

Flight period: late March and April (4 records).

S. maculata

Hosts: in Europe single records from *Euxoa obelisca* and the non-British *Ochropleura candelisequa* (D.&S.) (both Lepidoptera: Noctuidae). *E.obelisca* has a very restricted distribution in Britain and is not therefore an important host of this common species.

British distribution: England, Wales (Merioneth), S.Scotland (Edinburgh) and N. Scotland.

Flight period: late April to early June (at least 50 records).

S. mesnili

Hosts: unknown.

British distribution: S.England (Oxon, Berks, Herts, Cambs and Kent), Midlands, S.Scotland (Edinburgh) and N. Scotland.

Flight period: late April and May (16 records).

S. pauciseta

Hosts: in Britain: *Achlya flavicornis* (Lepidoptera: Thyatiridae) (NMW) and from galls of *Lipara* sp. (Diptera: Chloropidae) on reed (UMO). No European records.

British distribution: S.England (Chippenham Fen, Devil's Ditch and Wicken Fen in Cambs; Barton Mills and Dunwich in Suffolk; Horning Ferry and Fowlmere in Norfolk).

Flight period: late May to late July (18 records).

S. setosa

Hosts: in Europe single records from *Eupithecia succenturiata* (Lepidoptera: Geometridae) and *Allophyes oxyacanthae* (Lep.: Noctuidae).

British distribution: S.England (Berks and Sussex), Midlands (Salop), Wales, N.England (Cumberland) and N. Scotland.

Flight period: July and August (16 records).

S. variata

Hosts: in Britain *Mythimna litoralis* (Lepidoptera: Noctuidae) (3), a nocturnal feeder on Marram Grass.

British distribution: Wales (Glamorgan), N.England (Lancs) and N.Scotland (Inverness).

Flight period: 1.vi, 1.viii, 8-16.vii and 1-8.vii (the last 2 are eclosion dates)

Tribe Leskiini

The biology of most species in this tribe is very poorly known.

Genus **Aphria**

A. longirostris

Hosts: uncertain. In Britain *Hyloicus pinastri* (Lepidoptera: Sphingidae) (NHM). In Europe one record from *Sciota hostilis* (Lep.: Pyralidae). Emden (1954) states recorded from Agrotidae larvae.

British distribution: S.England (south of a Bristol-London line).

Habitat: heathland and downland (Emden, 1954). In Europe usually found in sandy areas (Herting, 1960).

Flight period: late May to late August (35 records).

Genus **Demoticus**

D. plebejus

Hosts: in Europe a questionable record from the non-British *Ammobiota festiva* (Hufnagel) (Lepidoptera: Arctiidae).

British distribution: S.England.

Habitat: downland (Emden, 1954).

Flight period: early June to early August (34 records).

Genus **Bithia**

B. modesta

Hosts: in Europe 2 records from unidentified Sesiidae (Lepidoptera).

British distribution: S.England (Barton and Farley Downs in Hants; 8 localities in Dorset).

Habitat: localities are chalk and limestone grassland, usually on the coast (Falk, in press). On the northern edge of its range (Tschorsnig, pers. comm.).

Flight period: late June to early August (20 records).

B. spreta
   Hosts: unknown.
   British distribution: S.England, Midlands (Hereford and Worcs), Wales and Ireland (Waterford, Kerry, Mayo and Wicklow*).
   Habitat: heathland and wastes (Emden, 1954).
   Flight period: early July to early September (at least 50 records).

Genus **Leskia**

L. aurea. Eggs are laid on the bark of infested trees and the first-instar larva searches for the host gallery. Overwinters as a second-instar larva in the host, pupating in the host gallery.
   Hosts: wood-boring Sesiidae larvae (Lepidoptera).
   British distribution: S.England (near Romsey in Hants, 1928 NHM).
   Flight period: in Europe July and August (Herting, 1960).

Genus **Solieria**

Only the males in this genus are distinguishable and even then only with difficulty. Literature references should therefore be treated with caution.

S. fenestrata
   Hosts: unknown.
   British distribution: S.England and Midlands (Worcs).
   Habitat: chalk wastes (Emden, 1954).
   Flight period: July and August (32 records).

S. inanis
   Hosts: in Britain a questionable old record from *Orthosia ?incerta* (Lepidoptera: Noctuidae) (Morley, 1906). References to *Spilosoma lutea* as a host are erroneous (based on misreading of Morley, 1906).
   British distribution: S.England, Midlands (Hereford and Worcs), Wales (Montgomery and ?Merioneth), N.England (Yorks*) and N.Scotland (Grantown in Moray; Inverness*).
   Habitat: heathland and downland (Emden, 1954).
   Flight period: July and August (24 records).

S. pacifica. Mechanism of contacting the host uncertain; the first-instar larva can crawl forwards but does not make searching movements with the anterior part of its body.
   Hosts: in Britain *Aglais urticae* (Lepidoptera: Nymphalidae) (BCM). In Europe a record from the non-British *Celypha rurestrana* (Duponchel) (Lep.: Tortricidae) which feeds on the root stock of *Heracleum*.
   British distribution: S.England, Midlands (Hereford and Worcs), Wales (Glamorgan) and N.England (Maltby in Yorks*).
   Habitat: chalk wastes (Emden, 1954).
   Flight period: early June to late August (at least 50 records).

S. vacua
   Hosts: in Europe a record from an unidentified Tortricidae larva (Lepidoptera).

British distribution: S.England, Midlands, Wales (Merioneth and Glamorgan), N.England (York).

Flight period: late July and August (33 records – including 3 in June).

Tribe Minthoini

Genus **Mintho**

M. rufiventris.

Hosts: in Britain the saprophagous *Orthopygia glaucinalis* (Lepidoptera: Pyralidae) (NHM). No European records.

British distribution: S.England (no records west of Berks-Sussex).

Habitat: localities are grassland and woodland (Falk, in press). Adults often found on the ground and on walls near buildings.

Flight period: late May to early September (48 records).

Tribe Microphthalmini

Where the biology is known, members of this tribe are all parasitoids of cockchafer larvae (Coleoptera: Scarabaeidae: Melolonthinae). The mobile first-instar larva contacts the host.

Genus **Dexiosoma**

D. caninum

Hosts: in Europe 3 questionable records from larvae of *Melolontha melolontha* (Coleoptera: Scarabaeidae).

British distribution: England, Wales, S.Scotland (Kirkcudbright), N.Scotland and Ireland (Cork, Wicklow, Waterford, Donegal*, Kerry, Mayo, Carlow, Down*, Antrim* and Fermanagh*).

Habitat: low vegetation in woodland (Herting, 1960; Chandler, pers. comm.).

Flight period: early July to late August (at least 50 records).

Sub-family Dexiinae

Tribe Dexiini

Parasitoids usually of large wood-boring or soil-dwelling Coleoptera larvae (possibly with the exception of *Trixa*). The mobile first-instar larva contacts the host.

Genus **Trixa**

The first-instar larvae of both species are described in Zuska (1962).

T. conspersa

Hosts: in contrast to the other Dexiini, this species may be a parasitoid of ground-dwelling Lepidoptera larvae; in Europe 2 records from the root-feeding *Hepialus lupulinus* (Hepialidae). In Britain a possibly erroneous record from *Operophtera fagata* (Lep.: Geometridae) (Audcent, 1942).

British distribution: England, Wales and N.Scotland.

Habitat: woodland margins and meadows (Herting, 1960).

Flight period: late May to early September (at least 50 records). 2 generations per year.

T. caerulescens

Hosts: unknown.

British distribution: S.England, Midlands (Hereford*).

Habitat: usually woodland margins (Chandler, pers. comm.)

Flight period: May and early June (at least 50 records). 1 generation per year.

Genus **Billaea**

B. irrorata. Overwinters as a larva within the host, killing it in the spring and pupating inside the gallery.

Hosts: mainly the stem-boring *Saperda populnea* (Coleoptera: Cerambycidae). In Britain 6 records from this species (1 on *Fraxinus excelsior* and 3 on *Populus tremula*). Also either this species or *Synanthedon flaviventris* (Lepidoptera: Sesiidae) (NHM), and from an unknown host within swellings on *Salix ?atrocinerea* (NHM). In Europe also *Paranthrene tabaniformes* (1) (Lep.: Sesiidae) and *Oberea* spp (5) (Cerambycidae).

British distribution: S.England, Midlands (Worcs) and S.Scotland (Edinburgh).

Habitat: most localities are woodland (Stubbs, pers. comm.)

Flight period: late May to late August (8 records). In Europe mid-July to the beginning of August (Herting, 1960).

Genus **Dinera**

D. carinifrons

Hosts: in Britain *Aphodius ater* larvae (Coleoptera: Scarabaeidae) (Pelham-Clinton, 1959), a dung feeder. No European records.

British distribution: England, Wales, and N.Scotland.

Flight period: early June to early September (at least 50 records).

D. grisescens

Hosts: in Britain *Harpalus ?affinis* larva (Coleoptera: Carabidae) (NHM). In Europe another record from this host (specific identity also uncertain).

British distribution: S.England, Midlands, Wales (Pembroke*) and N.England (Lancs).

Habitat: usually open areas. (Herting, 1960; Belshaw, 1992).

Flight period: late June to early August (at least 50 records). 1 generation per year.

Genus **Estheria**

As far as is known, all species in this genus are parasitoids of Scarabaeidae larvae (Coleoptera).

E. bohemani
Hosts: unknown.
British distribution: N.Scotland (sand dunes at Invernaver 28.vii.1972 (P.Chandler) and an unknown locality 1877 UMO both in Sutherland).

E. cristata
Hosts: in Britain *Phyllopertha horticola* (Coleoptera: Scarabaeidae) (NHM), a root-feeder. No European records.
British distribution: S.England, Midlands (Hereford and Worcs*), Wales (Glamorgan and Monmouth) and N.England (Cumberland).
Habitat: ?heathland (Day, 1948).
Flight period: July and August (at least 50 records).

Genus **Dexia**

D. rustica. Biology in Walker (1943). Between 1 and 6 larvae develop per host, overwintering within it in their second instar.
Hosts: root-feeding Melolonthinae larvae (Coleoptera: Scarabaeidae), chiefly *Melolontha* and *Amphimallon*. In Britain: *Melolontha melolontha, Amphimallon solstitialis* and *Phyllopertha horticola* (all Walker, 1943).
British distribution: S.England, Midlands (Hereford and Notts*), Wales and N.England (Yorks*).
Habitat: in Europe usually found in meadows, fields and woodland margins (Herting, 1960).
Flight period: early July to late August (at least 50 records). 1 generation per year.

D. vacua
Hosts: in Europe single records from *Serica brunnea* and (questionably) *Melolontha* sp. (both Coleoptera: Scarabaeidae: Melolonthinae).
British distribution: S.England, Midlands (Hereford and Cheshire), Wales, N.England and N.Scotland.
Flight period: late July and August (at least 50 records).

Genus **Prosena**

P. siberita. Overwinters as a larva within the host.
Hosts: in Europe single records from *Euchlora dubia* and *Euchlora* sp. (Coleoptera: Scarabaeidae: Rutelinae). In Japan other Scarabaeidae.
British distribution: S.England, Wales (Merioneth and Caernarvon), N.England, N.Scotland (Outer Hebrides) and Ireland (Kerry, Cork, Wicklow, Down* and Clare*).
Habitat: in Britain usually found in heathland and dunes (Chandler, pers. comm.). In Europe found in sandy areas (Herting, 1960).
Flight period: July and August (at least 50 records).

Tribe **Voriini**

Species in this tribe lay incubated eggs on larvae of Lepidoptera and sawflies (Hymenoptera: Symphyta).

Genus **Eriothrix**

E. prolixa
Hosts: in Britain *Pempelia obductella* (Lepidoptera: Pyralidae) (Hards, 1958 – in pods of *Silene cucubalus* but see Emmet, 1988; Uffen, 1961). In Europe a questionable record from the non-British *Pyrausta porphyralis* (D.&S.) (Pyralidae).
British distribution: S.England, Midlands (Hereford) and N.England (York*).
Habitat: swampy meadows (Herting, 1960). [Downland (Emden, 1954).]
Flight period: 2 periods – June and late July to early September (at least 50 records).

E. rufomaculata
Hosts: in Europe single records from the doubtfully British *Dendrolimus pini* (Lepidoptera: Lasiocampidae) and from the non-British *Ammobiota festiva* (Hufnagel) (Lep.: Arctiidae), both unlikely to be important hosts of this very common species (Herting, 1960).
British distribution: England, Wales, S.Scotland (Dumfries), N.Scotland and Ireland (Kilkenny, Kerry, Down*, Antrim*, Mayo, Dublin, Carlow and Galway).
Habitat: usually unimproved grassland (Chandler, pers. comm.).
Flight period: July and August (at least 50 records).

Genus **Campylocheta**

C. inepta. Overwinters as a puparium in the ground. May be gregarious.
Hosts: mainly Geometridae larvae but also several other families of Lepidoptera. In Britain: Geometridae – *Aplocera* sp. (NHM), *Ematurga atomaria* (3), also from an unidentified Geometridae (Ford, 1973). Noctuidae – *Anarta myrtilli* (Ford & Shaw, 1991).
British distribution: S.England, Midlands, Wales, N.England*, S.Scotland (Wigtown* and Dumfries*), N.Scotland and Ireland (Wicklow). The majority of records from N.Scotland.
Habitat: in Europe some association with cool montane areas with pine forests but the species is also found in warmer areas in central Europe and Spain (Tschorsnig, pers. comm.). Hosts are associated with heathland and moorland (Ford & Shaw, 1991).
Flight period: June and July (at least 50 records). 1 generation per year.

C. praecox. Gregarious, overwintering as a puparium in the ground.
Hosts: in Europe: *Thyatira batis* (1) (Lepidoptera; Thyatiridae), *Colotois pennaria* (2) and *Crocallis elinguaria* (2) (both Lep.; Geometridae).
British distribution: S.England, Midlands, Wales (Glamorgan) and N.Scotland (Loch Katrine in Perth).
Habitat: deciduous woodland (Herting, 1960).
Flight period: late March to early May (44 records).

Genus **Blepharomyia**

B. pagana. May be gregarious.
Hosts: mainly Geometridae larvae (Lepidoptera), but also several other families

of Lepidoptera, generally tree- or shrub-feeders. In Britain: Geometridae – *Agriopis leucophaearia* and *A.marginaria* both on *Quercus* (UMO), *Biston strataria* on *Quercus* (UMO), *Erannis defoliaria* (5 – inc. 3 on *Quercus*), *Eulithis populata* (Ford, 1973), *Opisthograptis luteolata* (3 – inc. 1 on *Crataegus*).

British distribution: S.England, Midlands (Worcs), Wales (Merioneth), N.England* and N.Scotland (Moray).

Habitat: usually deciduous woodland (Herting, 1960).

Flight period: early April to early June (at least 50 records).

B. piliceps

Hosts: in Europe mainly Lepidoptera larvae associated with shrubs in moorland: *Entephria caesiata* (1), *Perizoma didymata* (1), *Eulithis populata* (5), *Epirrita autumnata* (2), *Ematurga atomaria* (1), *Semiothisa brunneata* (1) (all Geometridae) and *Lithomoia solidaginis* (1) (Noctuidae).

British distribution: Midlands (Derby and Cheshire), N.England (Lancs) and N.Scotland.

Habitat: in Europe usually found in cool montane areas with pine forests (Tschorsnig, pers. comm.).

Flight period: May and June (20 records).

Genus **Ramonda**

R. latifrons.

Hosts: in Europe 2 records from *Mythimna ferrago* (Lepidoptera: Noctuidae), which usually feeds on grasses.

British distribution: S.England, Midlands (Hereford and Worcs) and Wales (Glamorgan).

Habitat: woodland (Emden, 1954).

Flight period: 2 periods – late May/early June and August/September (31 records).

R. prunaria

Hosts: Noctuidae larvae (Lepidoptera) associated with a range of plants.

British distribution: N.Scotland (Cairngorms in Inverness 26.vi.1933 NHM).

Habitat: not an alpine-boreal species; in Europe usually found in warm areas (Tschorsnig, pers. comm.).

Flight period: late April to early September, probably 2 generations per year (37 European records).

R. spathulata. Often gregarious.

Hosts: Noctuidae larvae (Lepidoptera), often low-feeding species in grassland (Ford & Shaw, 1991). In Britain: Noctuidae – *Xestia xanthographa* (9), *Actebia praecox* (Ford & Shaw, 1991), *Agrochola helvola* (UMO), *Apamea ?crenata* (Ford & Shaw, 1991), *Graphiphora augur* on *Rubus idaeus* (NHM), *Mythimna ?pallens* (Ford & Shaw, 1991), and *?Noctua pronuba* (Ford & Shaw, 1991). In Europe also recorded from *Agrotis* spp, *Cerastis rubricosa*, *Lycophotia porphyrea* and *Mythimna ferraga*.

British distribution: England, Wales and N.Scotland.

Habitat: woodland and heathland (Herting, 1960) but see 'Hosts'.

Flight period: early May to early October (at least 50 records, 50% in May). Probably 3 generations.

Genus **Periscepsia**

P. carbonaria
Hosts: in Europe: *Euxoa obelisca* (1), *Agrotis segetum* (1), *A.vestigialis* (1), *A.ipsilon* (1) and *Agrotis* sp. (2), all Noctuidae larvae (Lepidoptera) feeding at ground level.
British distribution: S.England (coastal counties only), Wales (Merioneth, Flint and Glamorgan) and Midlands (Lincs).
Habitat: sandy coastal and estuarine areas (Emden, 1954; Herting, 1960).
Flight period: early June to late September (at least 50 records).

Genus **Wagneria**

W. costata
Hosts: unknown. The record from *Hoplodrina blanda* is erroneous (Herting, 1960).
British distribution: S.England (Bere Wood* and Yellowham Wood in Dorset; Hale Purlieu in New Forest in Hants; Bookham Common in Surrey; Guestling in Sussex).
Habitat: localities are broadleaved woodland (Falk, in press).
Flight period: late May to early July (6 records). In Europe June and July (Herting, 1960).

W. gagatea
Hosts: in Europe single records from the following tree-feeding Lepidoptera larvae: *Orthosia cruda, O.cerasi* (both Noctuidae), *Operophtera brumata, Erannis defoliaria* (both Geometridae), *Dryomia ruficornis* (Notodontidae) and the non-British *Araschnia levana* (Nymphalidae). In Britain *Conistra vaccinii* (Noctuidae) on *Quercus* (UMO).
British distribution: S.England (Gloucs and Hunts) and Midlands (Hereford, Worcs and Warwick*).
Habitat: deciduous woodland (Falk, in press; Herting, 1960).
Flight period: late May to early July (11 records). In Europe May and June (Herting, 1960).

Genus **Athrycia**

Literature references to species in this genus should be treated with caution: the character given for separating *trepida* and *curvinervis* in Emden (1954) (where they are considered as forms) is unreliable, and *impressa* has only recently been recognised as occurring in Britain. At least the first 2 species appear to be facultatively gregarious.

A. curvinervis. Leaves the host to pupate in the ground, overwintering as a puparium.
Hosts: Noctuidae larvae (Lepidoptera), the majority of which feed nocturnally. In Britain: Noctuidae – *Euplexia lucipara* (Ford, 1976), *?Lacanobia contigua* (LM), *Lacanobia oleracea* (NHM; Ford, 1976), *Mamestra brassicae* (Ford, 1976), *Melanchra persicariae* (3), *Orthosia gracilis* on *Spiraea* (Hammond & Smith, 1955). Also recorded from most of these species in Europe (including *Lacanobia contigua*).

British distribution: S.England, Midlands (Staffs and Cheshire), Wales (Glamorgan), N. England (Lancs and York*) and Ireland (Cork).
Flight period: late June to late August (24 records).

A. impressa

Hosts: in Britain *Anarta myrtilli* (Lepidoptera: Noctuidae) (Ford & Shaw, 1991 – determined by B.Herting). In Europe single records from the above host and the non-British *Sideridis anapheles* (Nye) (Noctuidae) and *Rhyparia purpurata* (L.) (Lep.: Arctiidae).
British distribution: S.England (Linwood in Hants 30.viii.1971 NHM; Bagshot Heath in Surrey 20.viii.1934 NHM; Bexley in Kent 1971 NHM), probably wider (only recently recognised as occurring in Britain).

A. trepida

Hosts: Noctuidae larvae (Lepidoptera), chiefly Hadeninae. In Britain recorded from *Orthosia miniosa* (NHM) and *Orthosia gothica* (Ford & Shaw, 1991 – seen by author).
British distribution: S.England, Midlands, N.England (York*) and N.Scotland (3 females from Inverness, Argyll and Sutherland).
Flight period: late May to early July (at least 50 records).

Genus Voria

V. ruralis. Usually gregarious (up to 6), pupating in the empty host integument. Biology in Elsey & Rabb (1970), Grant & Shepard (1983) and Browning & Oatman (1984).
Hosts: chiefly Plusiinae larvae (Lepidoptera: Noctuidae) on herbs, but also a few other Lepidoptera. In Britain: Noctuidae – *Abrostola triplasia* (Ford & Shaw, 1991), *Autographa gamma* (7 + a questionable record on beet), *Autographa jota* (NHM), *Diachrysia chrysitis* (3 – inc. 1 on *Origanum vulgare* and 1 on *Mentha spicata*), also recorded from an unidentified Plusiinae on Mentha. Nymphalidae – *Vanessa atalanta* (BCM).
British distribution: England, Wales, S.Scotland (Kirkcudbright* and Midlothian) and N.Scotland (Ross and Moray).
Flight period: early June to early October (at least 50 records, 50% in late August/early September). More than 1 generation per year.

Genus Cyrtophleba

C. ruricola. At least in some cases gregarious.
Hosts: in Europe a number of Noctuidae (Lepidoptera), also a record from *Pachycnemia hippocastanaria* (Lep.: Geometridae).
British distribution: S.England (Witherington Downs in Wilts 11.vii.1972 NHM; Wimbledon Common in London 20.vi.1954 NHM; Wye Downs in Kent 6.vi.1936 NHM and UMO). Rarer in northern Europe but range extending to Norway, central Sweden and Finland (Tschorsnig, pers. comm.).
Flight period: in Europe late April to early August (Tschorsnig, pers. comm.).

Genus **Phyllomya**

P. volvulus

Hosts: Tenthredinidae sawfly larvae (Hymenoptera: Symphyta) on low plants. In Europe: *Macrophya albicincta* (2), *Pachyprotasis rapae* (2), *Tenthredo scrophulariae* (1), *Aglaostigma fulvipes* (1) and the non-British *A.nebulosa* (André) (1). No British records.

British distribution: England, Wales and N.Scotland (Perth*).

Habitat: woodland, usually running over low vegetation (Herting, 1960; Chandler, pers. comm.)

Flight period: late June to late August (at least 50 records). 1 generation per year.

Genus **Thelaira**

T. nigripes. One series overwintered as larvae within *Arctia caja*, producing an average of 5 adults per host individual (Hammond & Smith, 1955). In Europe the males are found on leaves and the females among ground vegetation (Herting, 1960).

Hosts: chiefly Arctiidae and other large Lepidoptera larvae on low plants. In Britain: Arctiidae – *Arctia villica* (Sperring, 1932), *Arctia caja* (3), *Spilosoma lubricipeda* (4), *S.lutea* (LM). Lasiocampidae – *Malacosoma neustria* (BCM). Sphingidae – *Deilephila porcellus* (NHM).

British distribution: England, Wales and Ireland (Galway, Down*, Derry*, Mayo, Waterford and Kerry).

Flight period: late June to early August (at least 50 records). 1 generation per year.

T. solivaga

Hosts: in Europe 6 records from Arctiidae larvae (Lepidoptera). In Britain *Arctia villica* (Arctiidae) (NHM).

British distribution: S.England (no records east of Bucks/Surrey), Midlands (Warwick*) and Wales (Glamorgan).

Flight period: early May to early September (at least 50 records).

Tribe Dufouriini

Where the biology is known, species in this tribe are parasitoids of Coleoptera, laying incubated eggs directly on the host.

Genus **Dufouria**

D. chalybeata. Appears to lay the egg on the host larva, overwintering as a larva within the adult and leaving it in the spring to pupate.

Hosts: *Cassida* spp (Chrysomelidae). In Britain: *C.rubiginosa* (2 – NHM) and *C.viridis* (NHM).

British distribution: S.England, (southern) Midlands, Wales and N.England (Lancs*).

Flight period: early June to early July (at least 50 records). 1 generation per year.

D. nigrita

Hosts: unknown. References to *Cassida nobilis* and *C.vittata* as hosts are erroneous (Herting, 1960).

British distribution: S.England, (southern) Midlands, Wales (Merioneth and Brecon) and N.Scotland (Callander in Perth).

Flight period: June and July (at least 50 records).

Genus **Rondania**

R. fasciata

Hosts: a single record from *Phyllobius argentatus* (Coleoptera: Curculionidae) (NHM). Other members of the genus are also parasitoids of adult weevils. *R.dimidiata* pushes its ovipositor (which is similar to that of *R.fasciata*) into the oesophagus of the feeding host.

British distribution: S.England, Midlands, N.England (York) and N.Scotland (Aviemore in Inverness*).

Habitat: localities are woodland (Falk, in press).

Flight period: May and June (38 records).

Genus **Microsoma**

M. exigua. Lays incubated eggs on the outside of its adult host. The larva apparently enters the host between the head and thorax, later pupating in the host integument (Berry & Parker, 1950).

Hosts: chiefly *Sitona* spp (Coleoptera: Curculionidae), also single records from *Hypera postica* and the non-British *Polydrusus inustus* Germ. (both Curculionidae). No British records.

British distribution: S.England (Berks, Hants, Surrey, Sussex and Kent), Midlands (Hereford and Worcs).

Flight period: late May to early August (28 records).

Genus **Freraea**

F. gagatea. May be gregarious. The tip of the ovipositor is weakly sclerotised, indicating it probably does not penetrate the host integument (Herting, 1960).

Hosts: adult Coleoptera. In Britain *Harpalus tardus* (Carabidae) (NHM). In Europe: *Harpalus rufipes* (3), *Amara aulica* (1) (both Carabidae) and questionably the non-British *Agrilus viridis* (1) (Buprestidae).

British distribution: S.England (Wilts, Berks and Suffolk) and N.Scotland (Pitlochry in Perth*).

Habitat: localities are heathland and grassland (Falk, in press).

Flight period: 10.vi.; 22.vi.; 17.vii.; 9.viii.; 20.viii. In Europe end of June and July (Herting 1960).

Subfamily Phasiinae

The British Phasiinae, as far as is known, are all parasitoids of heteropteran bugs (Hemiptera), laying (with the exception of *Redtenbacheria insignis*) unincubated eggs. These are either laid on, or inserted into, the (usually adult) host. In species employing the latter strategy, the females have highly modified terminalia to assist with oviposition. The major work on the biology of the group in Europe and North Africa is Dupuis (1963).

Tribe Eutherini

Genus **Redtenbacheria**

R. insignis. Appears to lay incubated eggs (Herting, 1966).

Hosts: unknown. The record from *Lymantria monacha* (Lepidoptera: Lymantriidae) is considered unlikely by Herting (1966).

British distribution: S.England (Parkend in Gloucs 29.vii.1945 NHM; Plympton in Devon 16.vii.1934 UMO; Lyndhurst in Hants 4.vii.1894 NHM and 26.vi.1897 UMO).

Habitat: localities indicate woodland (Falk, in press).

Flight period: in Europe June and July (Herting, 1960).

Tribe Phasiini

Genus **Subclytia**

S. rotundiventris

Hosts: in Britain: *Elasmostethus interstinctus* on *Betula* (Allen, 1966) and *Elasmucha grisea* (NHM – on grass) (both Hemiptera: Acanthosomatidae). The latter species is usually found on *Betula*. In Europe further single records from these species, *Piezodorus lituratus* (Hemiptera: Pentatomidae) and *Cyphostethus tristriatus* (Acanthosomatidae).

British distribution: S.England (Wilts, Hants, Berks, Herts*, Surrey, Sussex and Bucks) and Midlands (Sandwell Valley*).

Habitat: localities are woodland, heathland and chalk grassland (Falk, in press).

Flight period: early May to early September (14 records).

Genus **Gymnosoma**

G. nitens. The egg is laid on the ventral surface of the (usually adult) host's abdomen. Overwinters as a larva within the host.

Hosts: in Europe: *Sciocoris cursitans* (5) and the non-British *S.helferi* Fb. (1) (Hemiptera: Pentatomidae).

British distribution: S.England (Happy Valley near Boxhill in Surrey 8.vii.1956 NHM).

Flight period: in Europe 2 generations per year – end of May/June and July to September (Herting, 1960).

G. rotundatum. The egg is laid on the fourth- or fifth-instar nymph or adult host. Overwinters as a larva within the host, leaving to pupate in the ground.

Hosts: *Palomena* spp (Hemiptera: Pentatomidae), also many old and questionable records from other Pentatomidae. No British records.

British distribution: S.England (London and Kent*; numerous records from Surrey and Sussex) and Ireland (Cork and Kerry*).

Habitat: dry sandy areas on downland and heathland with isolated shrubs (Falk, in press).

Flight period: early June to early September (14 records).

Genus **Cistogaster**

C. globosa. The egg is laid on the dorsal surface of the host's abdomen. Leaves the host to pupate in the ground.

Hosts: in Europe *Aelia* (10) (Pentatomidae), mostly *A.acuminata*.

British distribution: S.England (Portsdown in Hants; Cothill NNR in Berks; White Downs in Surrey*).

Habitat: localities indicate downland (Falk, in press).

Flight period: in Europe late May to late August. Possibly 2 generations per year (Tschorsnig, pers. comm.).

Genus **Opesia**

O. cana. The structure of the egg and the female terminalia indicate that the egg is laid on the host.

Hosts: unknown.

British distribution: S.England.

Habitat: localities indicate deciduous woodland (Falk, in press).

Flight period: May and early June (12 records).

Genus **Phasia**

Females in this genus possess a strongly sclerotised, posteriorly directed, piercing structure (derived from sternite 8). The egg is therefore presumably inserted into the host.

P. hemiptera

Hosts: in Europe single records from *Palomena prasina, Pentatoma rufipes* and the non-British *P.metallifera* (Mtsch.) (all Hemiptera: Pentatomidae). No British records.

British distribution: S.England (south of a Bistol-London line), Midlands (Hereford and Worcs*), Wales (Merioneth), N.England*, S.Scotland (Kirkcudbright*, Dumfries*, Ayr*, Stirling* and Fife*), N.Scotland (Perth) and Ireland (Clare, Waterford, Wexford*, Kerry*, Dublin*, Wicklow*, Carlow* and Down*).

Habitat: in Scotland found in stable, floristically rich, habitats, e.g. old meadows and gorge woodlands previously managed by coppice rotation (Bayne, 1987). In England found in woodland margins and old meadows (Chandler, pers. comm.).

Flight period: early May to early August (at least 50 records),

P. obesa

Hosts: heteropteran bugs (Hemiptera). In Britain *Neottiglossa pusilla* (Pentatomidae) (Allen, 1963). In Europe single records from *Zicrona coerulea* (Pentatomidae), *Leptopterna dolabrata* (Miridae) and the non-British *Sehirus melanopterus* (H.-S.) (Cydnidae), these hosts all associated with grassland. Also questionable single European records (based on immature stages) from *Lygus pratensis* (Miridae), *Beosus maritimus* (Lygaeidae) and *Myrmus miriformis* (Rhopalidae).

British distribution: S.England, Midlands (Hereford and Worcs), N.England (York*), N.Scotland and Ireland (Kerry and Waterford).

Habitat: found in grassland, scrub and woodland margins (Chandler, pers. comm.)

Flight period: early June to late September, peaking August (at least 50 records).

P. pusilla

Hosts: in Europe a range of heteropteran bugs (Hemiptera) – Cydnidae, Lygaeidae, Berytinidae and Cimicidae. In Britain: *Stygnocoris fuligineus* and *S.pedestris* (Lygaeidae) (Eyles, 1962).

British distribution: S.England, Midlands and N.Scotland (Inverness*).

Habitat: woodland margins and grassland (Chandler, pers. comm.).

Flight period: 2 periods – late May/early June and early July to early September (at least 50 records).

Tribe Catharosiini

Genus **Litophasia**

L. hyalipennis. Placed among the Rhinophoridae in Emden (1954) and unlikely to be recognised as a tachinid owing to its reduced subscutellum.

Hosts: unknown.

British distribution: S.England (Guestling near Hastings in Sussex 1887*; Kingsnorth and Northfleet in Kent (Clemons, 1992)).

Flight period: August (2 British and 5 European records).

Tribe Leucostomatini

The females of the British species in this tribe have characteristic terminalia. Sternite 8 forms an anteriorly-directed piercing structure, and tergite 6 forms a pair of posteriorly-directed forceps at the tip of the abdomen. The operation of these structures is not known but they are presumably used to insert the egg into the host.

Genus **Dionaea**

D. aurifrons. Apparent male swarming behaviour observed by Fonseca (1949).

Hosts: in Europe a record from *Dicranocephalus agilis* (Hemiptera: Stenocephalidae) and in Japan a record from the non-British *Riptortus clavatus* (Thb.) (Hemiptera: Alydidae).

British distribution: S.England (Putsborough and Braunton Burrows in Devon).

Habitat: localities indicate sand dunes (Falk, in press).

Flight period: late May to late August (6 British and 10 European records).

Genus **Leucostoma**

L. simplex

Hosts: in Europe a record from *Aptus mirmicoides* (Hemiptera: Nabidae).

British distribution: S.England (Hurn 4.viii.1932 UMO, Aldridge Hill 19.vii.1954 NHM and Latchmore 13.vii.1971 NHM in Hants; Slough in Bucks 22.viii.1936 NHM; Sussex*; Norfolk*).

Flight period: in Europe June to August (Herting, 1960).

Genus **Labigastera**

L. forcipata. First recorded as British in Spooner (1974).

Hosts: in Europe single records from *Dicranocephalus agilis* (Hemiptera: Stenocephalidae) and *Enoplops scapha* (Hemiptera: Coreidae).

British distribution: S.England (West Looe in Cornwall 11.vii.1972 and 26.vi.1984 NHM).

Flight period: in Europe June to August (Herting, 1960).

Genus **Cinochira**

C. atra. Attack both nymphs and adults, emerging from the anus or genital aperture of the adult to pupate. Overwinter as mature larvae occupying the entire body cavity of the host (Eyles, 1962).

Hosts: in Britain: *Drymus brunneus*, *D.sylvaticus*, *Scolopostethus decoratus* and *S.thomsoni* (Hemiptera: Lygaeidae) (Eyles, 1962). In Europe a record from *Eremocoris plebejus* (Lygaeidae).

British distribution: S.England, Midlands (Hereford and Cheshire*), N.England (Yorks*) and Ireland (Dublin*).

Habitat: low vegetation in woodland (Chandler, pers. comm.).

Flight period: early June to early September (28 records).

Tribe **Cylindromyiini**

The females is this tribe have characteristic terminalia. Abdominal segments 6 and 7 are folded under the anterior abdominal segments. Tergite 7 ends in two hook-shaped projections. Sternite 10 forms a piercing structure which can be projected through the hooks of tergite 7. This piercer however is small and weakly sclerotised and is presumably capable of penetrating only soft membranes. The operation of these structures is not known but they are presumably used to insert the egg into the host.

Genus **Lophosia**

L. fasciata

Hosts: in Britain *Acanthosoma haemorrhoidale* (Hemiptera: Acanthosomatidae) (Allen, 1987). In Europe a record from *Aelia acuminata* (Hemiptera: Pentatomidae).

British distribution: S.England and Ireland (Kerry and Cork).

Habitat: localities are downland, coastal grassland and dry woodland (Falk, in press).

Flight period: late July and early August (29 records).

Genus **Cylindromyia**

C. brassicaria. Solitary, overwintering as a second-instar larva in the host and pupating in the soil.

Hosts: Palaearctic records from *Dolycoris* spp (mostly *D.baccarum*), also single questionable records from *Holcostethus vernalis* and *Palomena prasina* (all Hemiptera: Pentatomidae). Other records from Pentatomidae and Scutelleridae are probably erroneous.

British distribution: S.England (Lizard Peninsula in Cornwall; Hambledon Hill in Dorset/Wilts; Farley in Wilts) and Ireland (Glengariff in Cork).

Habitat: English localities are grassland and scrub (Falk, in press).

Flight period: July and August (6 records). In Europe June to August (Herting, 1960).

C. interrupta
Hosts: unknown.

British distribution: S.England.

Flight period: early June to late August (34 records).

Genus **Hemyda**

H. vittata. A total of 17 records from Britain: the earliest is 1956 and the others are all post-1970. Probably therefore a recent arrival. Overwinters as a larva in the host.

Hosts: in Europe: the non-British *Arma custos* (F.) (3) and questionably *Troilus luridus* (1) (both Hemiptera: Pentatomidae).

British distribution: S.England (Berks*, Herts, Hants* and Sussex).

Habitat: localities indicate broadleaved and mixed woodland (Falk, in press).

Flight period: early May to late August, probably 2 generations per year (9 British plus 22 European records).

Genus **Phania**

References to Carabidae (Coleoptera) as hosts of a non-British species in this genus are probably erroneous (Herting, 1960).

P. funesta
Hosts: in Europe a record from *Legnotus limbosus* (Hemiptera: Cydnidae).

British distribution: S.England and Midlands*.

Flight period: late May to late August (at least 50 records).

P. thoracica
Hosts: unknown.

British distribution: S.England (Abbots Wood in Hants; Guestling in Sussex; Blean in Kent) and Midlands (Mains Wood in Hereford).

Habitat: localities indicate woodland (Falk, in press).

Flight period: July (3 records).

Check List

The following list includes all generic and specific names which have been used in the British literature since Wainwright (1928). In addition, all specific names in Mesnil (1944-1975 and 1980) are included. Older names used in Britain may be found in Kloet & Hincks (1975), and a complete list of those used in the Palaearctic may be found in Herting (1984).

The classification follows Herting (1984). The nomenclature also follows this work except where it is in contravention of the I.C.Z.N. code. In addition, two more recent changes have been adopted (Andersen, 1988; O'Hara, 1989).

The incorrect names used in the earlier tachinid Handbook (Emden, 1954) are marked with an asterisk (*).

EXORISTINAE
EXORISTINI
EXORISTA Meigen, 1803
    TACHINA of authors*, not Meigen, 1803
    TRICHOLYGA Rondani, 1856
    PODOTACHINA Brauer & Bergenstamm, 1891
fasciata (Fallén, 1820)
larvarum (Linnaeus, 1758)
glossatorum (Rondani, 1859)
    baranoffi (Wainwright, 1933)*
grandis (Zetterstedt, 1844)
    sorbillans of authors*, not Wiedemann, 1830
mimula (Meigen, 1824)
    minor (Wainwright, 1932)
    nigricans (Egger, 1861)*
    erucarum (Rondani, 1859)
rustica (Fallén, 1810)
    simulans of authors, not Meigen, 1824
tubulosa Herting, 1967
    erucarum of authors*, not Rondani, 1859

CHETOGENA Rondani, 1856
    CHAETOGENA, variant spelling
    STOMATOMYIA Brauer & Bergenstamm, 1889*
acuminata Rondani, 1859

DIPLOSTICHUS Brauer & Bergenstamm, 1889
janithrix (Hartig, 1838)

PARASETIGENA Brauer & Bergenstamm, 1891
silvestris (Robineau-Desvoidy, 1863)
    agilis of authors, not Robineau-Desvoidy, 1830

PHOROCERA Robineau-Desvoidy, 1830
assimilis (Fallén, 1810)
obscura (Fallén, 1810)
    vernalis Robineau-Desvoidy, 1830
    caesifrons Macquart, 1850

BESSA Robineau-Desvoidy, 1863
    PTYCHOMYIA Brauer & Bergenstamm, 1889*
parallela (Meigen, 1824)
    fugax (Rondani, 1861)
selecta (Meigen, 1824)

BLONDELIINI
BELIDA Robineau-Desvoidy, 1863
    APOROTACHINA Meade, 1894*
angelicae (Meigen, 1824)

MEIGENIA Robineau-Desvoidy, 1830
dorsalis (Meigen, 1824)
    pilosa Baranov, 1926*
    discolor (Zetterstedt, 1838)
majuscula (Rondani, 1859)
mutabilis (Fallén, 1810)
    floralis of authors, not Fallén, 1810
    bisignata (Meigen, 1824)*

ZAIRA Robineau-Desvoidy, 1830
    BIOMYA Rondani, 1856
    VIVIANIA Rondani, 1861*
cinerea (Fallén, 1810)

GASTROLEPTA Rondani, 1862
    MEDORIA of authors*, not Robineau-Desvoidy, 1830
anthracina (Meigen, 1826)

MEDINA Robineau-Desvoidy, 1830
    DEGEERIA Meigen, 1838*
collaris (Fallén, 1820)
luctuosa (Meigen, 1830)
    funebris of authors, not Meigen, 1830
separata (Meigen, 1824)

POLICHETA Rondani, 1856
    PERICHETA Rondani, 1859
    PERICHAETA, variant spelling*
unicolor (Fallén, 1820)
    funebris (Zetterstedt, 1838)

LEIOPHORA Robineau-Desvoidy, 1863
    ARRHINOMYIA of authors*, not
        Brauer & Bergenstamm, 1889
    APATELIA Stein, 1924, preoccupied
innoxia (Meigen, 1824)

ADMONTIA Brauer & Bergenstamm, 1889
    TRICHOPAREIA Brauer & Bergen-
        stamm, 1889*
    TRICHOPARIA, variant spelling
blanda (Fallén, 1820)
grandicornis (Zetterstedt, 1849)
    amica of authors*, not Meigen, 1838
    podomyia of authors, not Brauer &
        Bergenstamm, 1889
maculisquama (Zetterstedt, 1859)
    seria of authors*, not Meigen, 1824
seria (Meigen, 1824)
    decorata (Zetterstedt, 1849)*

OSWALDIA Robineau-Desvoidy, 1863
muscaria (Fallén, 1810)
    sordidisquama (Zetterstedt, 1844)

HEMIMACQUARTIA Brauer & Bergen-
    stamm, 1893
paradoxa Brauer & Bergenstamm, 1893

LIGERIA Robineau-Desvoidy, 1863
    ANACHAETOPSIS Brauer & Bergen-
        stamm, 1889*
angusticornis (Loew, 1847)
    angustifrons: mispelling in Herting, 1960
    zetterstedti (Ringdahl, 1945)*
    ocypterina of authors, not Zetterstedt,
        1838.

BLONDELIA Robineau-Desvoidy, 1830
nigripes (Fallén, 1810)

COMPSILURA Bouché, 1834
concinnata (Meigen, 1824)

VIBRISSINA Rondani, 1861
    MICROVIBRISSINA Villeneuve, 1911*
debilitata (Pandellé, 1896)
    villeneuvei (Wainwright, 1940)*
    muscaria of authors, not Fallén, 1810

WINTHEMIINI
RHAPHIOCHAETA Brauer & Bergen-
    stamm, 1889
breviseta (Zetterstedt, 1838)

SMIDTIA Robineau-Desvoidy, 1830
conspersa (Meigen, 1824)

TIMAVIA Robineau-Desvoidy, 1863

NEMOSTURMIA Townsend, 1926*
    CHETOLIGA of authors, not Rondani,
        1856
amoena (Meigen, 1824)

WINTHEMIA Robineau-Desvoidy, 1830
cruentata (Rondani, 1859)
quadripustulata (Fabricius, 1794)
variegata (Meigen, 1824)

NEMORILLA Rondani, 1856
floralis (Fallén, 1810)
    maculosa of authors, not Meigen, 1824
    notabilis (Meigen, 1824)

ERYCIINI
APLOMYA Robineau-Desvoidy, 1830
    APLOMYIA, variant spelling
confinis (Fallén, 1820)

PHEBELLIA Robineau-Desvoidy, 1846
glauca (Meigen, 1824)
glirina (Rondani, 1859)
stulta (Zetterstedt, 1844)
    quadriseta (Villeneuve, 1910)
    cotei of authors*, not Grilat, 1915
vicina (Wainwright, 1940)
villica (Zetterstedt, 1838)
    aestivalis Robineau-Desvoidy, 1846
    ingens (Brauer & Bergenstamm, 1891)*
    vicina of authors*, not Wainwright, 1940

NILEA Robineau-Desvoidy, 1863
hortulana (Meigen, 1824)

TLEPHUSA Robineau-Desvoidy, 1863
cincinna (Rondani, 1859)
    diligens of authors*, not Zetterstedt,
        1844
    honesta Robineau-Desvoidy, 1863

EPICAMPOCERA Macquart, 1849
succincta (Meigen, 1824)

PHRYXE Robineau-Desvoidy, 1830
heraclei (Meigen, 1824)
    latilobata Wainwright, 1940*
magnicornis (Zetterstedt, 1838)
    longicauda Wainwright, 1940*
nemea (Meigen, 1824)
vulgaris (Fallén, 1810)

BACTROMYIA Brauer & Bergenstamm,
    1891
aurulenta (Meigen, 1824)

PSEUDOPERICHAETA Brauer & Bergen-
    stamm, 1889

nigrolineata (Walker, 1853)
  insidiosa (Robineau-Desvoidy, 1863)
  roseanae Brauer & Bergenstamm, 1891*

LYDELLA Robineau-Desvoidy, 1830
  PARAPHOROCERA Brauer & Bergen-
    stamm, 1889*
grisescens Robineau-Desvoidy, 1830
  senilis of authors*, not Meigen, 1838
stabulans (Meigen, 1824)

CADURCIELLA Villeneuve, 1927
tritaeniata (Rondani, 1859)

DRINO Robineau-Desvoidy, 1863
  PHORCIDA of authors*, not Robineau-
    Desvoidy, 1863
lota (Meigen, 1824)

HUEBNERIA Robineau-Desvoidy, 1847
affinis (Fallén, 1810)

CARCELIA Robineau-Desvoidy, 1830
  EURYCLEA Robineau-Desvoidy, 1863
  PELMATOMYIA Brauer and Bergen-
    stamm, 1889
atricosta Herting, 1961
gnava (Meigen, 1824)
  excavata (Zetterstedt, 1844)
lucorum (Meigen, 1824)
  comata (Rondani, 1859)
puberula Mesnil, 1941
rasa (Macquart, 1849)
  amphion Robineau-Desvoidy 1863
tibialis (Robineau-Desvoidy, 1863)
  phalaenaria of authors, not Rondani, 1859

SENOMETOPIA Macquart, 1834
  EUCARCELIA Baranov, 1934
excisa (Fallén, 1820)
intermedia (Herting, 1960)
pollinosa (Mesnil, 1941)
  obesa of authors, not Zetterstedt, 1859
  rutilla of authors*, not Rondani, 1859

THECOCARCELIA Townsend, 1933
acutangulata (Macquart, 1850)
  incedens (Rondani, 1861)

ERYCIA Robineau-Desvoidy, 1830
furibunda (Zetterstedt, 1844)
  festinans of authors, not Meigen, 1824
  fatua of authors, not Meigen, 1824
  cinerea of authors*, not Robineau-
    Desvoidy, 1863

XYLOTACHINA Brauer & Bergenstamm,
  1891
diluta (Meigen, 1824)

ligniperdae Brauer & Bergenstamm, 1891*

TOWNSENDIELLOMYIA Baranov, 1932
nidicola (Townsend, 1908)

GONIINI
PLATYMYA Robineau-Desvoidy, 1830
  PLATYMYIA, variant spelling
fimbriata (Meigen, 1824)
  nemestrina (Meigen, 1824)

EUMEA Robineau-Desvoidy, 1863
linearicornis (Zetterstedt, 1844)
  westermanni (Zetterstedt, 1844)*
  mitis of authors, not Meigen, 1824

MYXEXORISTOPS Townsend, 1911
blondeli (Robineau-Desvoidy, 1830)
  pexops (Brauer & Bergenstamm, 1891)
stolida (Stein, 1924)
  pexops of authors*, not Brauer & Bergen-
    stamm, 1891

ZENILLIA Robineau-Desvoidy, 1830
libatrix (Panzer, 1798)
  macrops (Brauer & Bergenstamm, 1891)
  ciligera of authors, not Robineau-
    Desvoidy, 1830

CLEMELIS Robineau-Desvoidy, 1863
pullata (Meigen, 1824)

PALES Robineau-Desvoidy, 1830
pavida (Meigen, 1824)

PHRYNO Robineau-Desvoidy, 1830
vetula (Meigen, 1824)

CYZENIS Robineau-Desvoidy, 1863
  MONOCHAETA Brauer & Bergen-
    stamm, 1889*
albicans (Fallén, 1810)

ERYCILLA Mesnil, 1957
ferruginea (Meigen, 1824)
  rutila of authors, not Meigen, 1824

OCYTATA Gistel, 1848
  RACODINEURA Rondani, 1861
  RHACODINEURA, variant spelling*
pallipes (Fallén, 1820)
  antiqua (Meigen, 1824)

EURYSTHAEA Robineau-Desvoidy, 1863
  DISCOCHAETA Brauer & Bergen-
    stamm, 1889*
scutellaris (Robineau-Desvoidy, 1848)
  evonymellae of authors*, not Ratzeburg,
    1844

ERYNNIA Robineau-Desvoidy, 1830
ocypterata (Fallén, 1810)
  nitida Robineau-Desvoidy, 1830*

ELODIA Robineau-Desvoidy, 1863
ambulatoria (Meigen, 1824)
  convexifrons (Zetterstedt, 1844)*
  cloacellae (Kramer, 1910)
morio (Fallén, 1820)
  tragica (Meigen, 1824)*

HEBIA Robineau-Desvoidy, 1830
flavipes Robineau-Desvoidy, 1830

FRONTINA Meigen, 1838
laeta (Meigen, 1824)

THELYMORPHA Brauer & Bergenstamm,
  1889
  ISTOCHETA of authors, not Rondani,
  1859
  HISTOCHAETA Bezzi, 1907*
marmorata (Fabricius, 1805)

BRACHICHETA Rondani, 1861
  BRACHYCHAETA, variant spelling*
strigata (Meigen, 1824)

GONIA Meigen, 1803
  SALMACIA Meigen, 1800, suppressed
capitata (DeGeer, 1776)
divisa Meigen, 1826
ornata Meigen, 1826
picea (Robineau-Desvoidy, 1830)
  fasciata Meigen, 1826, preoccupied
  sicula of authors*, not Robineau-
  Desvoidy, 1830

TACHININAE
TACHININI
TACHINA Meigen, 1803
  LARVAEVORA Meigen, 1800, suppressed
  ECHINOMYA Latreille, 1804
  ECHINOMYIA, variant spelling*
fera (Linnaeus, 1761)
grossa (Linnaeus, 1758)
subgenus SERVILLIA Robineau-
  Desvoidy, 1830*
lurida (Fabricius, 1781)
ursina (Meigen, 1824)

NOWICKIA Wachtl, 1894
ferox (Panzer, 1809)

PELETERIA Robineau-Desvoidy, 1830
  PELETIERIA, variant spelling*
rubescens (Robineau-Desvoidy, 1830)
  nigricornis (Meigen, 1838)*

GERMARIA Robineau-Desvoidy, 1830
  ATRACTOCHAETA Brauer & Bergenstamm, 1889*
angustata (Zetterstedt, 1844)
ruficeps (Fallén, 1820)

NEMORAEINI
NEMORAEA Robineau-Desvoidy, 1830
pellucida (Meigen, 1824)

LINNAEMYIINI
LINNAEMYA Robineau-Desvoidy 1830
  LINNAEMYIA, variant spelling*
  MICROPALPIS Macquart, 1834
  MICROPALPUS, variant spelling
comta (Fallén, 1810)
  compta, variant spelling
rossica Zimin, 1954
  haemorrhoidalis of authors*, not Fallén,
  1810
tessellans (Robineau-Desvoidy, 1830)
  pudica (Rondani, 1859)*
vulpina (Fallén, 1810)

CHRYSOCOSMIUS Bezzi, 1907
  CHRYSOSOMOPSIS Townsend, 1916*
auratus (Fallén, 1820)

LYDINA Robineau-Desvoidy, 1830
aenea (Meigen, 1824)

LYPHA Robineau-Desvoidy, 1830
  MICRONYCHIA Brauer & Bergen-
  stamm, 1889*
  EVERSMANIA of authors, not
  Robineau-Desvoidy, 1863
  EVERSMANNIA, variant spelling
dubia (Fallén, 1810)
ruficauda (Zetterstedt, 1838)

ERNESTIINI
ERNESTIA Robineau-Desvoidy, 1830
  MERIANIA Robineau-Desvoidy, 1830*
  PANZERIA Robineau-Desvoidy, 1830
laevigata (Meigen, 1838)
  nielseni (Villeneuve, 1921)*
puparum (Fabricius, 1794)
rudis (Fallén, 1810)
vagans (Meigen, 1824)

APPENDICIA Stein, 1924
truncata (Zetterstedt, 1838)

FAUSTA Robineau-Desvoidy, 1830
nemorum (Meigen, 1824)

EURITHIA Robineau-Desvoidy, 1844
  VARICHAETA Speiser, 1903
anthophila (Robineau-Desvoidy, 1830)
  radicum of authors*, not Linnaeus, 1758
caesia (Fallén, 1810)
connivens (Zetterstedt, 1844)
consobrina (Meigen, 1824)
intermedia (Zetterstedt, 1844)
  conjugata (Zetterstedt, 1852)*
vivida (Zetterstedt, 1838)

HYALURGUS Brauer & Bergenstamm, 1893
lucidus (Meigen, 1824)

GYMNOCHETA Robineau-Desvoidy, 1830
  GYMNOCHAETA, variant spelling*
viridis (Fallén, 1810)

ZOPHOMYIA Macquart, 1835
temula (Scopoli, 1763)

CLEONICE Robineau-Desvoidy, 1863
callida (Meigen, 1824)

LOEWIA Egger, 1856
  FORTISIA Rondani, 1861
foeda (Meigen, 1824)
phaeoptera (Meigen, 1824)
submetallica (Macquart, 1855)
  petiolata of authors*, not Robineau-Desvoidy, 1830

ELOCERIA Robineau-Desvoidy, 1863
  HELOCERA, variant spelling*
delecta (Meigen, 1824)

PELATACHININI
PELATACHINA Meade, 1894
tibialis (Fallén, 1810)

MACQUARTIINI
MACQUARTIA Robineau-Desvoidy, 1830
dispar (Fallén, 1820)
  flavipes (Meigen, 1824)
grisea (Fallén, 1810)
nudigena Mesnil, 1972
  buccalis of authors*, not Robineau-Desvoidy, 1830
praefica (Meigen, 1824)
pubiceps (Zetterstedt, 1845)
  nubilis (Rondani, 1862)*
tenebricosa (Meigen, 1824)
  nitida (Zetterstedt, 1838)
  chalconota of authors*, not Meigen, 1824
viridana Robineau-Desvoidy, 1863
  flavipes of authors*, not Meigen 1824

ANTHOMYIOPSIS Townsend, 1916
  PTILOPSINA Villeneuve, 1920*
nigrisquamata (Zetterstedt, 1838)
  nigrisquama, variant spelling
  nitens (Zetterstedt, 1852)*

TRIARTHRIINI
TRIARTHRIA Stephens, 1829
  BIGONICHETA Rondani, 1845
  DIGONOCHAETA, variant spelling*
setipennis (Fallén, 1810)
  spinipennis (Meigen, 1824)*

NEAERINI
NEAERA Robineau-Desvoidy, 1830
laticornis (Meigen, 1824)
  albicollis (Meigen, 1824)*

PHYTOMYPTERA Rondani, 1845
  ELFIA Robineau-Desvoidy, 1850
  CRASPEDOTHRIX Brauer & Bergenstamm, 1893*
cingulata (Robineau-Desvoidy, 1830)
  zonella of authors*, not Zetterstedt, 1844
nigrina (Meigen, 1824)
  nitidiventris Rondani, 1845*

GRAPHOGASTER Rondani, 1868
brunnescens Villeneuve, 1907

SIPHONINI
GONIOCERA Brauer & Bergenstamm, 1891
versicolor (Fallén, 1820)

ENTOMOPHAGA Lioy, 1864
exoleta (Meigen, 1824)
nigrohalterata (Villeneuve, 1921)

CEROMYA Robineau-Desvoidy, 1830
  CEROMYIA, variant spelling
  STENOPARIA Stein, 1924, preoccupied*
bicolor (Meigen, 1824)
monstrosicornis (Stein, 1924)
silacea (Meigen, 1824)

ACTIA Robineau-Desvoidy, 1830
  THRYPTOCERA Macquart, 1834
crassicornis (Meigen, 1824)
infantula (Zetterstedt, 1844)
  antennalis (Rondani, 1859)*
lamia (Meigen, 1838)
  frontalis (Macquart, 1845)*
maksymovi Mesnil, 1952
nudibasis Stein, 1924
pilipennis (Fallén, 1810)

PERIBAEA Robineau-Desvoidy, 1863
fissicornis (Strobl, 1909)

CERANTHIA Robineau-Desvoidy, 1830
abdominalis (Robineau-Desvoidy, 1830)
  anomala (Zetterstedt, 1849)*
  lichtwardtiana (Villeneuve, 1931)

APHANTORHAPHOPSIS Townsend, 1926
  ASIPHONA Mesnil, 1954
verralli (Wainwright, 1928)

SIPHONA Meigen, 1803
  CROCUTA Meigen, 1800, suppressed
  BUCENTES Latreille, 1809
boreata Mesnil, 1960
collini Mesnil, 1960
cristata (Fabricius, 1805)
geniculata (DeGeer, 1776)
ingerae Andersen, 1982
maculata Staeger, 1849
mesnili Andersen, 1982
pauciseta Rondani, 1865
  delicatula Mesnil, 1960
setosa Mesnil, 1960
variata Andersen, 1982

LESKIINI
APHRIA Robineau-Desvoidy, 1830
longirostris (Meigen, 1824)

DEMOTICUS Macquart, 1854
plebejus (Fallén, 1810)
  plebeius, variant spelling*

BITHIA Robineau-Desvoidy, 1863
  RHINOTACHINA Brauer & Bergen-
    stamm, 1889*
modesta (Meigen, 1824)
spreta (Meigen, 1824)

LESKIA Robineau-Desvoidy, 1830
aurea (Fallén, 1820)

SOLIERIA Robineau-Desvoidy, 1848
  MYOBIA Robineau-Desvoidy, 1830,
    preoccupied
  ANTHOICA Rondani, 1861
fenestrata (Meigen, 1824)
  fuscana Robineau-Desvoidy, 1848*
inanis (Fallén, 1810)
pacifica (Meigen, 1824)
  tibialis (von Roser, 1840)*
vacua (Rondani, 1861)

MINTHOINI
MINTHO Robineau-Desvoidy, 1830

rufiventris (Fallén, 1816)
  lacera Rondani, 1847*

MICROPHTHALMINI
DEXIOSOMA Rondani, 1856
caninum (Fabricius, 1781)

DEXIINAE
DEXIINI
TRIXA Meigen, 1824
caerulescens Meigen, 1824
  coerulescens: misprint in Emden, 1954
  alpina of authors, not Meigen, 1824
conspersa (Harris, 1776)
  oestroidea (Robineau-Desvoidy, 1830)*

BILLAEA Robineau-Desvoidy, 1830
  ATROPIDOMYIA Brauer & Bergen-
    stamm, 1889*
irrorata (Meigen, 1826)

DINERA Robineau-Desvoidy, 1830
  PHOROSTOMA Robineau-Desvoidy,
    1830
  MYOCERA Robineau-Desvoidy, 1830
  MYIOCERA, variant spelling
carinifrons (Fallén, 1816)
grisescens (Fallén, 1816)

ESTHERIA Robineau-Desvoidy, 1830
bohemani (Rondani, 1862)
cristata (Meigen, 1826)

DEXIA Meigen, 1826
rustica (Fabricius, 1775)
vacua (Fallén, 1816)

PROSENA LePeletier & Serville, 1828
  CALIRRHOE Meigen, 1800, suppressed
siberita (Fabricius, 1775)
  sybarita, variant spelling
  luculliana Rondani, 1861*

VORIINI
ERIOTHRIX Meigen, 1803
  RHYNCHISTA Rondani, 1861*
prolixa (Meigen, 1824)
rufomaculata (DeGeer, 1776)
  dimano (Harris, 1780)*
  monochaeta Wainwright, 1928*

CAMPYLOCHETA Rondani, 1859
  CAMPYLOCHAETA, variant spelling*
  ELPE Robineau-Desvoidy, 1863*
  HYPOCHAETA Brauer & Bergenstamm
    1889

inepta (Meigen, 1824)
praecox (Meigen, 1824)
  obscura of authors, not Fallén, 1810

BLEPHAROMYIA Brauer & Bergenstamm, 1889
pagana (Meigen, 1824)
  amplicornis (Zetterstedt, 1844)*
piliceps (Zetterstedt, 1859)
  collini Wainwright, 1928*

RAMONDA Robineau-Desvoidy, 1863
latifrons (Zetterstedt, 1844)
prunaria (Rondani, 1861)
  cunctans of authors, not Meigen, 1824
spathulata (Fallén, 1820)
  lentis (Meigen, 1824)*

PERISCEPSIA Gistel, 1848
carbonaria (Panzer, 1798)
  nigrans (Meigen, 1826)

WAGNERIA Robineau-Desvoidy, 1830
costata (Fallén, 1815)
gagatea Robineau-Desvoidy, 1830
  succincta (Meigen, 1838)*

ATHRYCIA Robineau-Desvoidy, 1830
curvinervis (Zetterstedt, 1844)
impressa (Wulp, 1869)
trepida (Meigen, 1824)

VORIA Robineau-Desvoidy, 1830
  PLAGIA Meigen, 1838
ruralis (Fallén, 1810)

CYRTOPHLEBA Rondani, 1856
ruricola (Meigen, 1824)

PHYLLOMYA Robineau-Desvoidy, 1830
  PHYLLOMYIA, variant spelling*
volvulus (Fabricius, 1794)

THELAIRA Robineau-Desvoidy, 1830
nigripes (Fabricius, 1794)
  leucozona of authors, not Panzer, 1809
solivaga (Harris, 1780)

DUFOURIINI
DUFOURIA Robineau-Desvoidy, 1830
  MINELLA of authors*, not Robineau-Desvoidy, 1830
chalybeata (Meigen, 1824)
nigrita (Fallén, 1810)

RONDANIA Robineau-Desvoidy, 1850
fasciata (Macquart, 1834)

MICROSOMA Macquart, 1855
  CAMPOGASTER Rondani, 1856
  SYNTOMOGASTER Egger, 1860*
exigua (Meigen, 1824)

FRERAEA Robineau-Desvoidy, 1830
gagatea Robineau-Desvoidy, 1830
  albipennis (Zetterstedt, 1838)*

PHASIINAE
EUTHERINI
REDTENBACHERIA Schiner, 1861
insignis Egger, 1861

PHASIINI
SUBCLYTIA Pandellé, 1894
rotundiventris (Fallén, 1820)

GYMNOSOMA Meigen, 1803
  RHODOGYNE Meigen, 1800, suppressed
  STYLOGYMNOMYIA Brauer & Bergenstamm, 1891
nitens Meigen, 1824
rotundatum (Linnaeus, 1758)

CISTOGASTER Latreille, 1829
globosa (Fabricius, 1775)

OPESIA Robineau-Desvoidy, 1863
  XYSTA of authors*, not Meigen, 1824
cana (Meigen, 1824)

PHASIA Latreille, 1804
  ALOPHORA Robineau-Desvoidy, 1830*
  ALLOPHORA, variant spelling
  HYALOMYA Robineau-Desvoidy, 1830
hemiptera (Fabricius, 1794)
obesa (Fabricius, 1798)
pusilla Meigen, 1824

CATHAROSIINI
LITOPHASIA Girschner, 1887
hyalipennis (Fallén, 1815)

LEUCOSTOMATINI
DIONAEA Robineau-Desvoidy, 1830
aurifrons (Meigen, 1824)

LEUCOSTOMA Meigen, 1803
simplex (Fallén, 1815)

LABIGASTERA Macquart, 1834
  LABIGASTER, variant spelling
forcipata (Meigen, 1824)

CINOCHIRA Zetterstedt, 1845
atra Zetterstedt, 1845

CYLINDROMYIINI
LOPHOSIA Meigen, 1824
  LOPHROSIA, variant spelling
fasciata Meigen, 1824

CYLINDROMYIA Meigen, 1803
  OCYPTERA Latreille, 1804*
brassicaria (Fabricius, 1775)
interrupta (Meigen, 1824)

HEMYDA Robineau-Desvoidy, 1830
  EVIBRISSA Rondani, 1861
vittata (Meigen, 1824)

PHANIA Meigen, 1824
  WEBERIA of authors*, not Robineau-
    Desvoidy, 1830
funesta (Meigen, 1824)
  pseudofunesta (Villeneuve, 1931)*
  curvicauda of authors, not Fallén, 1820
thoracica Meigen, 1824

Fig. 301. Female *Microsoma exigua* ovipositing on an adult weevil (Coleoptera: Curculioni-dae: *Sitona*). Drawn from observations in Berry & Parker (1950).

127

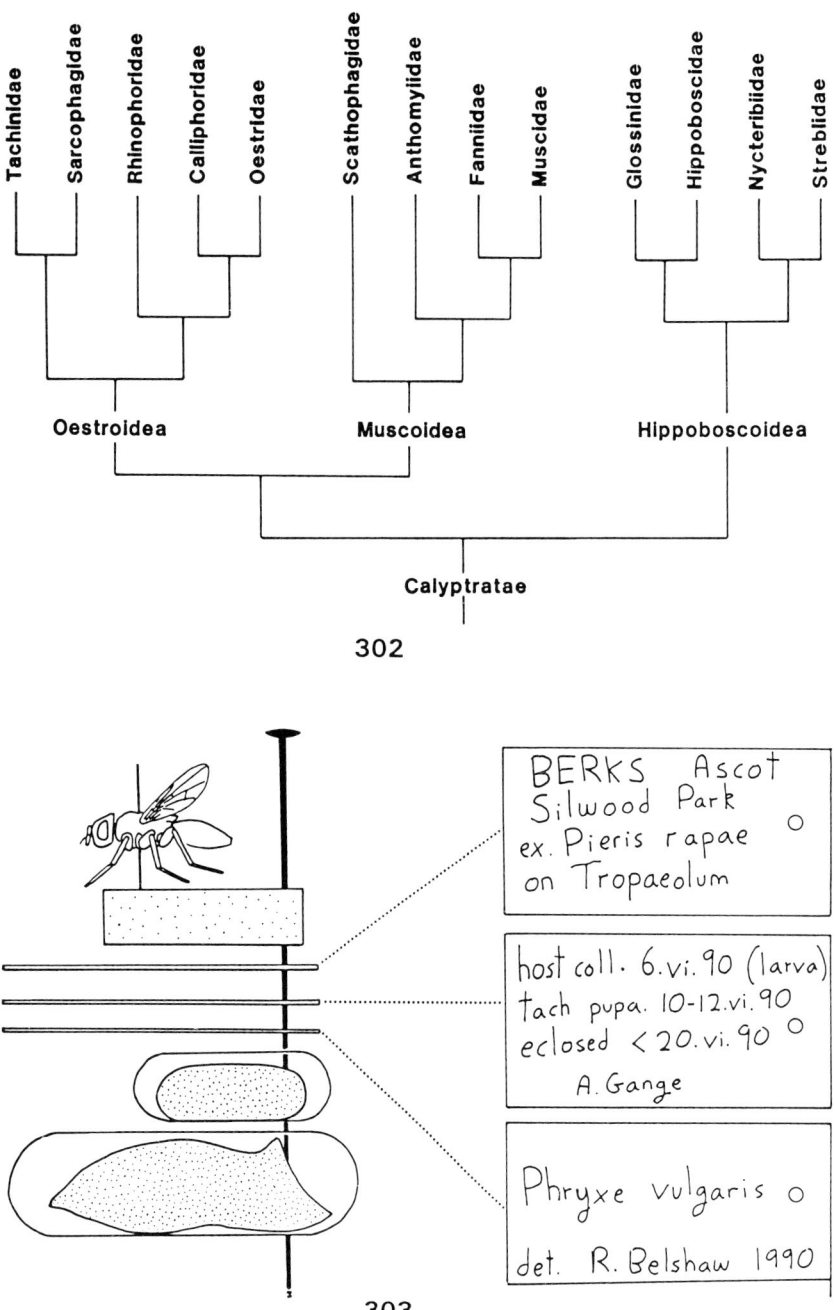

302

303

Fig. 302. Cladogram showing the most likely evolutionary relationships between the Tachinidae and other families of Calyptratae (Diptera). Following McAlpine (1989) except for the relationships within the Oestroidea which follow Pape (1992).

Fig. 303. Example of a pinned Tachinidae specimen together with collection and rearing data.

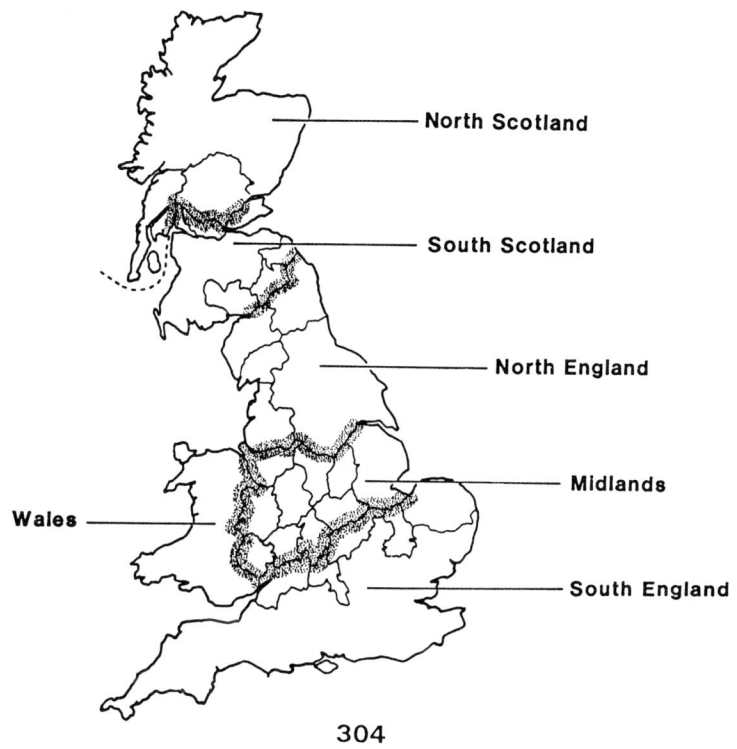

North Scotland

South Scotland

North England

Midlands

Wales

South England

304

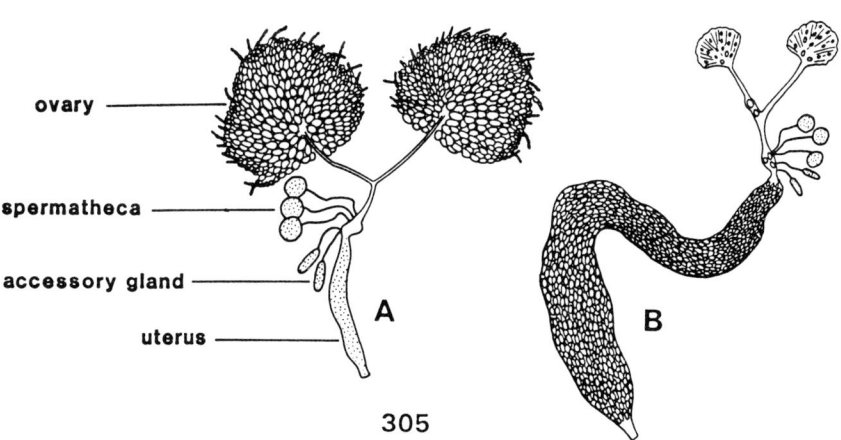

ovary

spermatheca

accessory gland

uterus

A

B

305

Fig. 304. Geographical regions of Great Britain used in the Species biology section.
Fig. 305. Female reproductive system of *Zenillia libatrix*: (a) unfertilised. (b) gravid (showing eggs accumulating in the uterus). After Ferrar (1987).

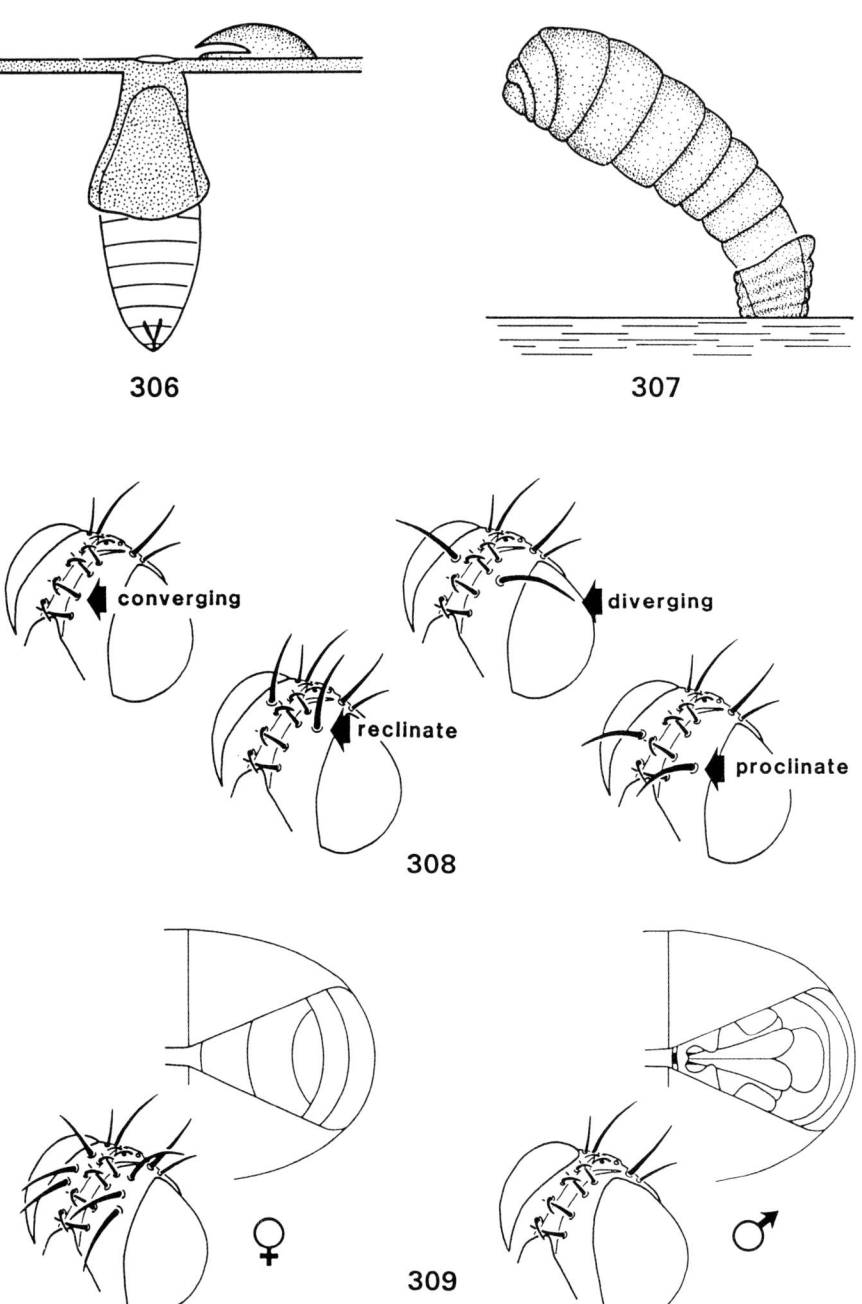

Fig. 306. First-instar larva of *Exorista larvarum* in its respiratory funnel, with adjacent empty egg case. After Ferrar (1987).
Fig. 307. First-instar larva of *Ernestia rudis*. After Ferrar (1987).
Fig. 308. Nomenclature of bristle orientation on the head.
Fig. 309. Common sexual dimorphism in tachinids. In head and abdomen (seen from below).

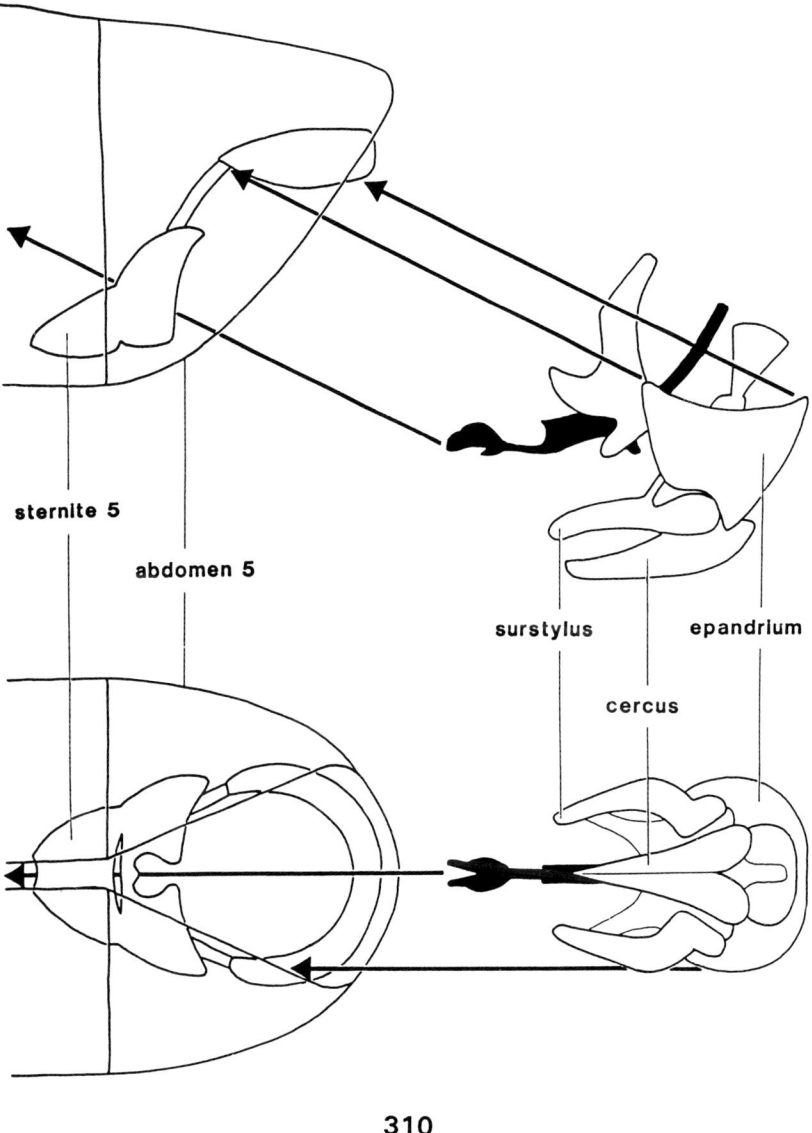

sternite 5

abdomen 5

surstylus   **epandrium**

cercus

310

Fig. 310. Schematic view of a typical male abdomen showing the position of the terminalia within the abdomen (seen from the side and from below).

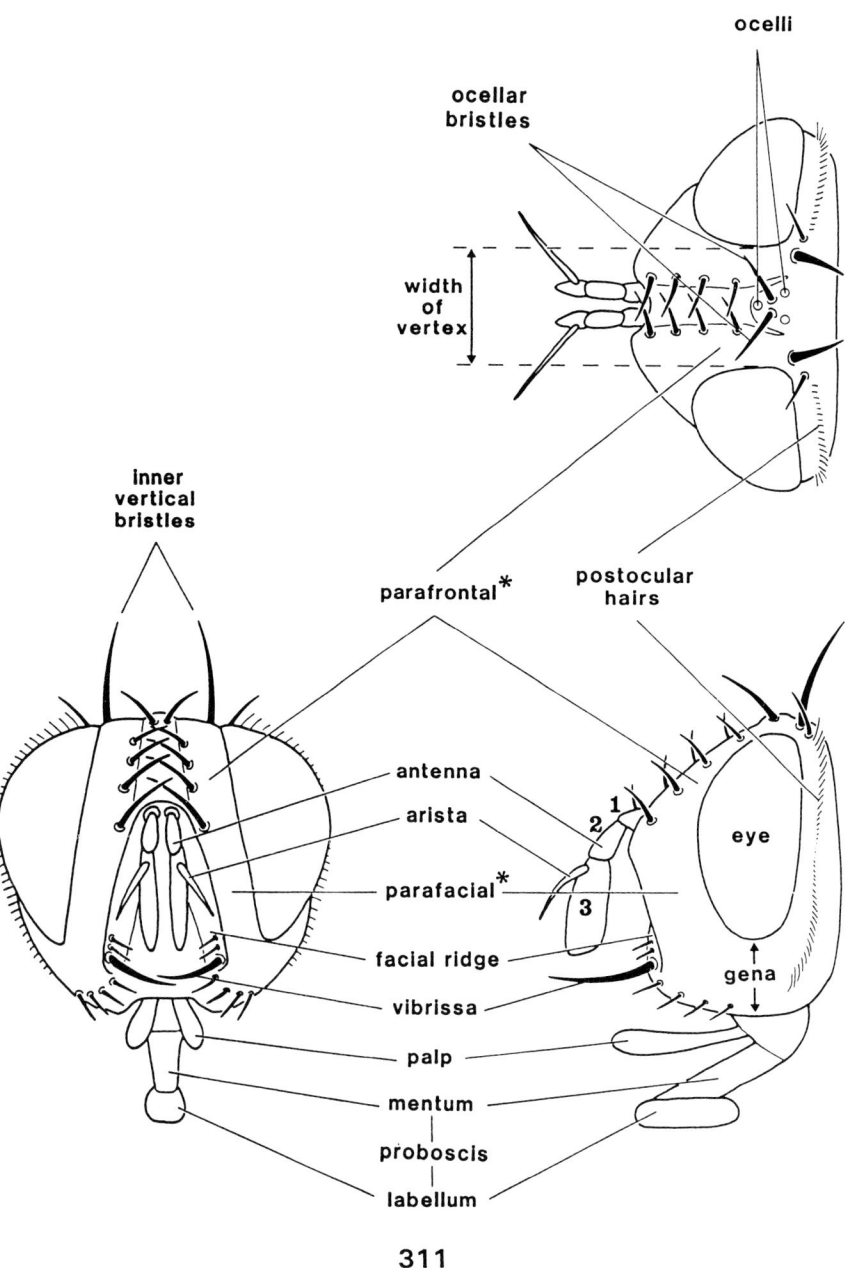

ocelli

ocellar
bristles

width
of
vertex

inner
vertical
bristles

parafrontal*

postocular
hairs

antenna

arista

eye

parafacial*

facial ridge

gena

vibrissa

palp

mentum

proboscis

labellum

311

Fig. 311. Head seen from the front, from above and from the side. * In the key the division between the parafrontal and parafacial areas is taken as being the position of the most ventral bristle **except** when it refers to the presence of bristles on the parafacial area.

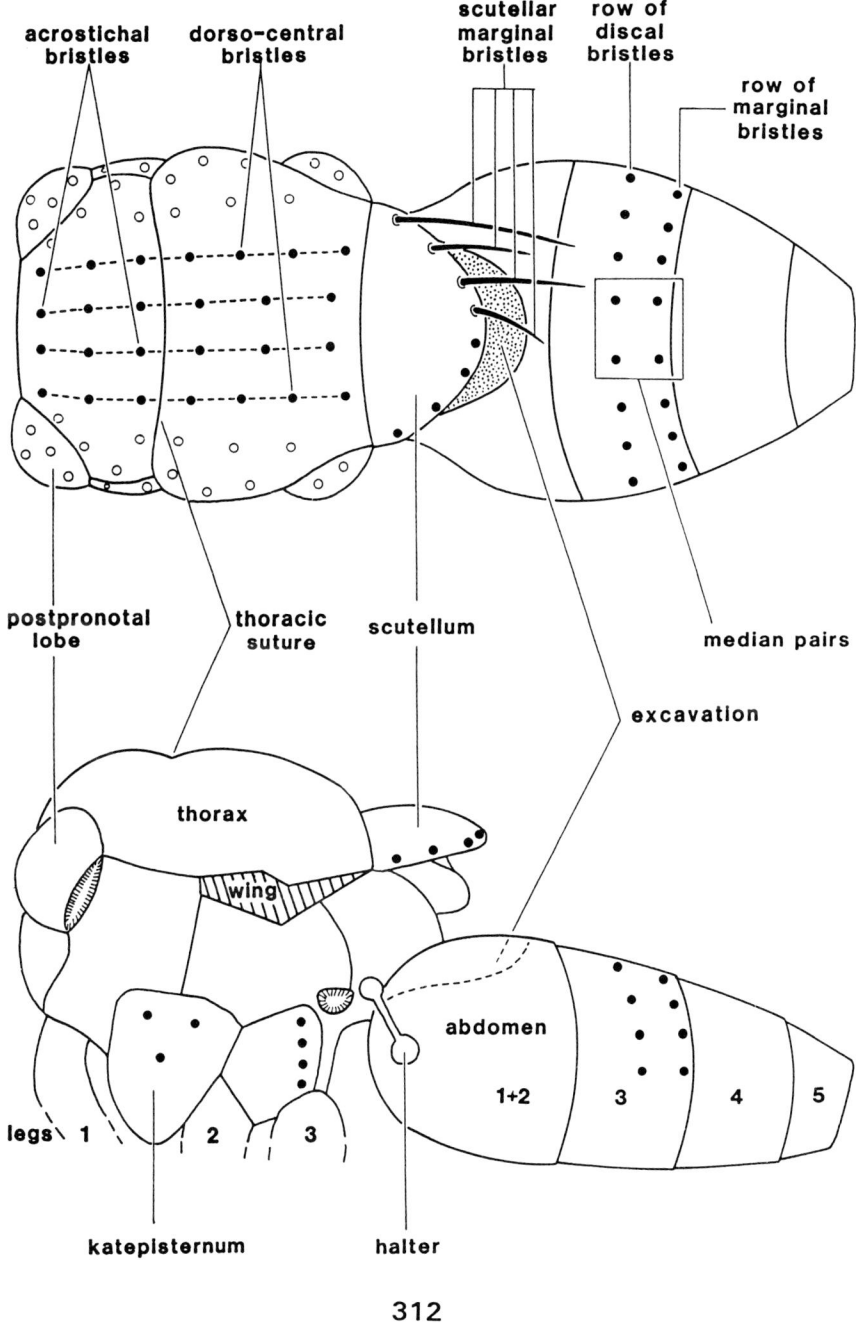

acrostichal bristles

dorso-central bristles

scutellar marginal bristles

row of discal bristles

row of marginal bristles

postpronotal lobe

thoracic suture

scutellum

median pairs

excavation

thorax

wing

abdomen

1+2

3

4

5

legs 1 2 3

katepisternum

halter

312

Fig. 312. Thorax and abdomen seen from above and from the side.

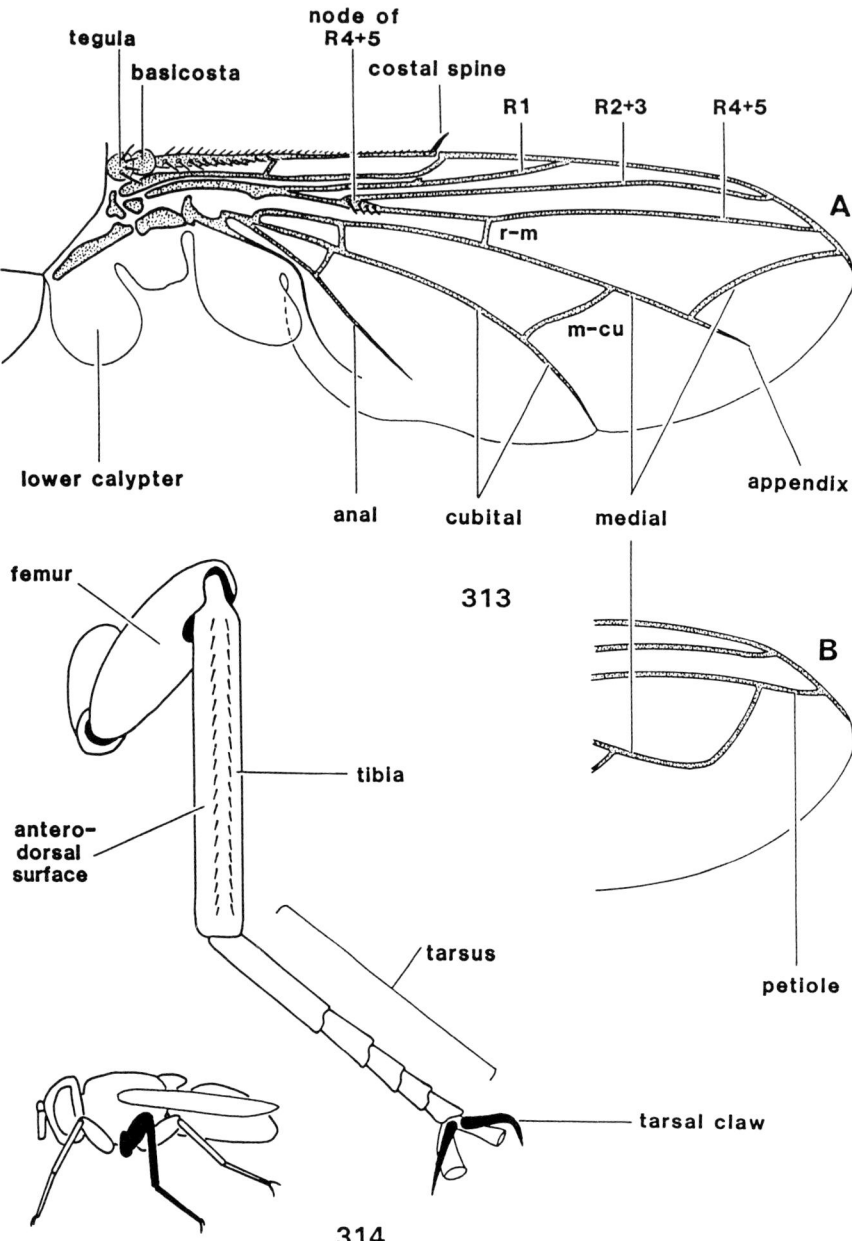

Fig. 313. Wing seen from above. The 2 main types of venation are shown: (a) the medial vein reaching the wing margin independently; (b) the medial vein joining vein R4+5 to form a petiole.

Fig. 314. Leg seen from the side. Its orientation in relation to the body is also shown, N.B. the same orientation is used for all 3 pairs of legs.

134

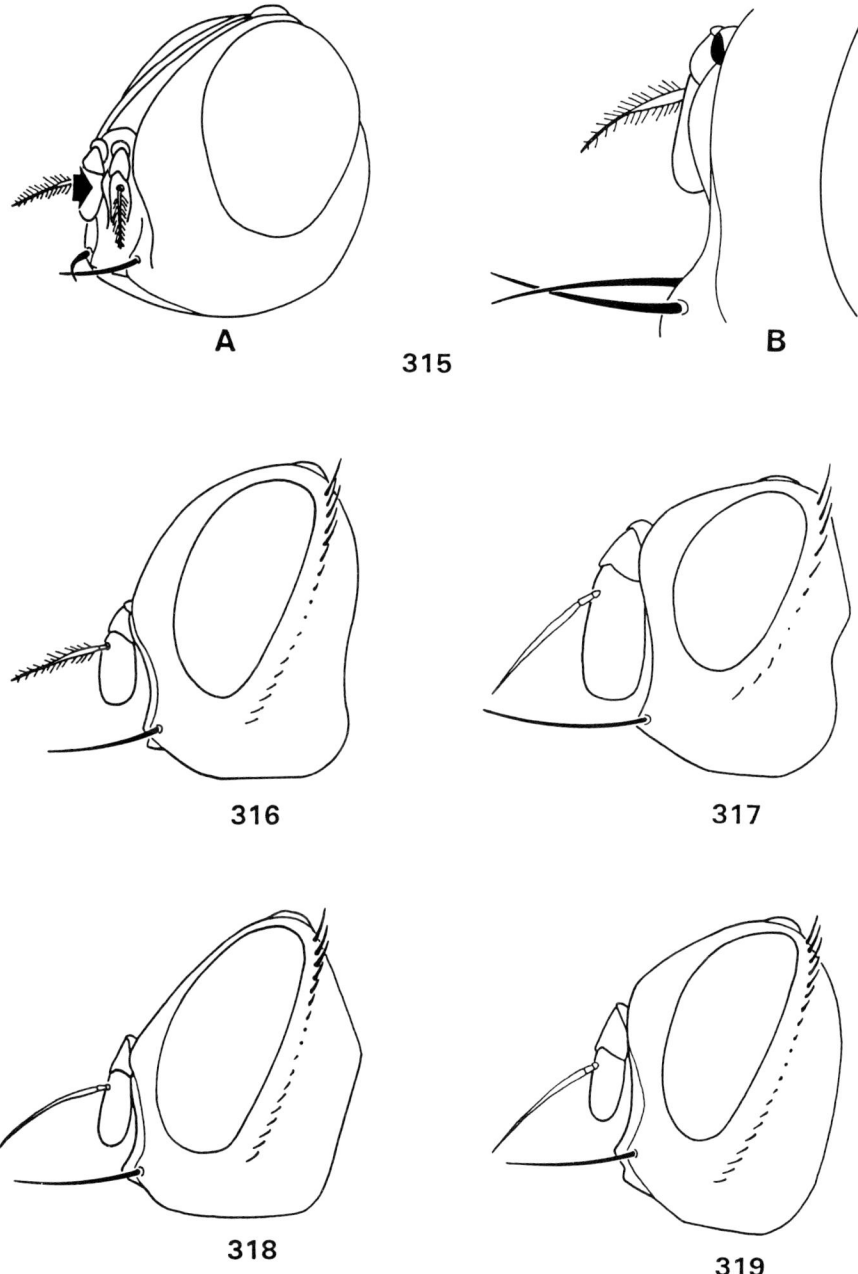

Figs 315–319. Head seen from the side.
315. *Dinera grisescens*: (a) showing ridge between the antennae; (b) a close-up of the ridge seen from the side with the left antenna removed.
316. *Billaea irrorata*
317. *Lydina aenea* (male).
318. *Dufouria chalybeata* (female).
319. *Macquartia praefica* (female).

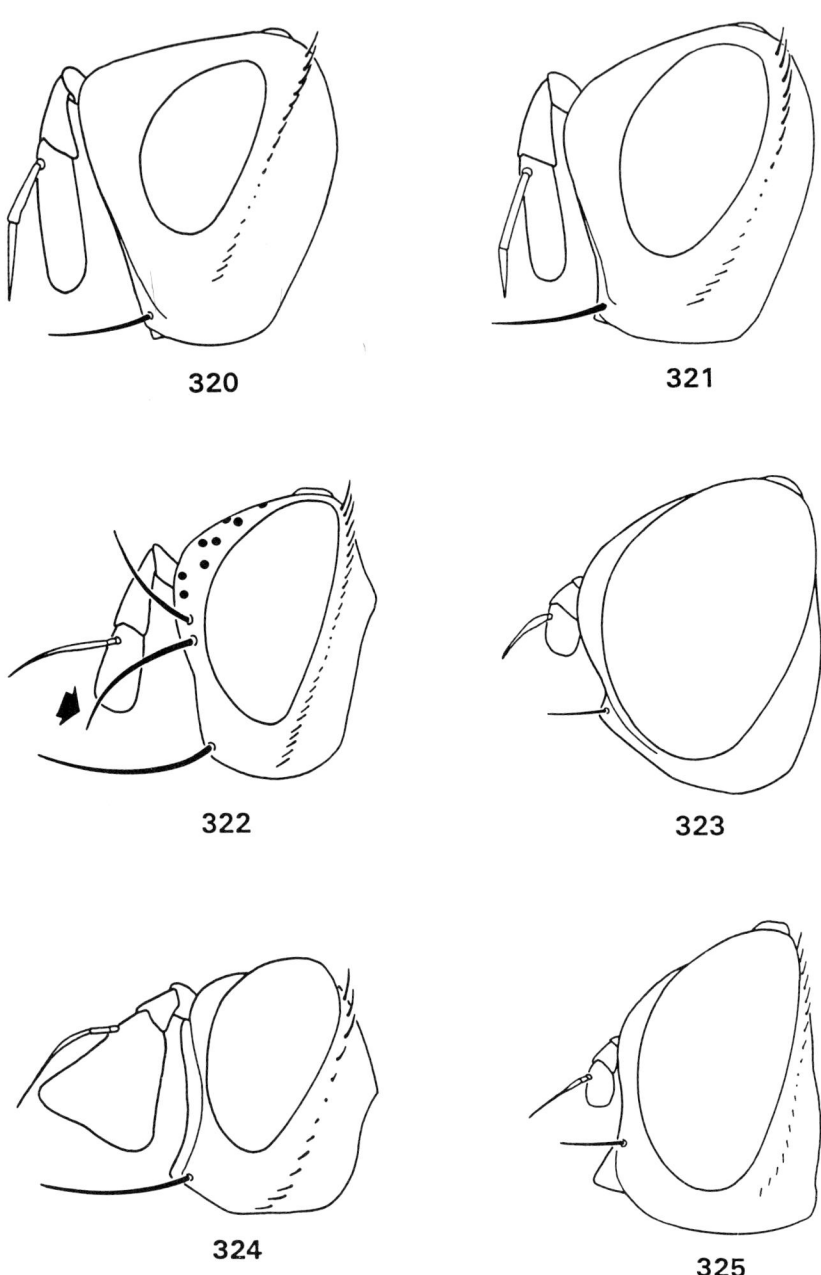

Figs 320–325. Head seen from the side.
320. *Germaria ruficeps* (female).
322. *Voria ruralis* (male and female).
324. Male *Lophosia fasciata*.

321. *Germaria angustata* (female).
323. Male *Freraea gagatea*.
325. *Opesia cana* (female).

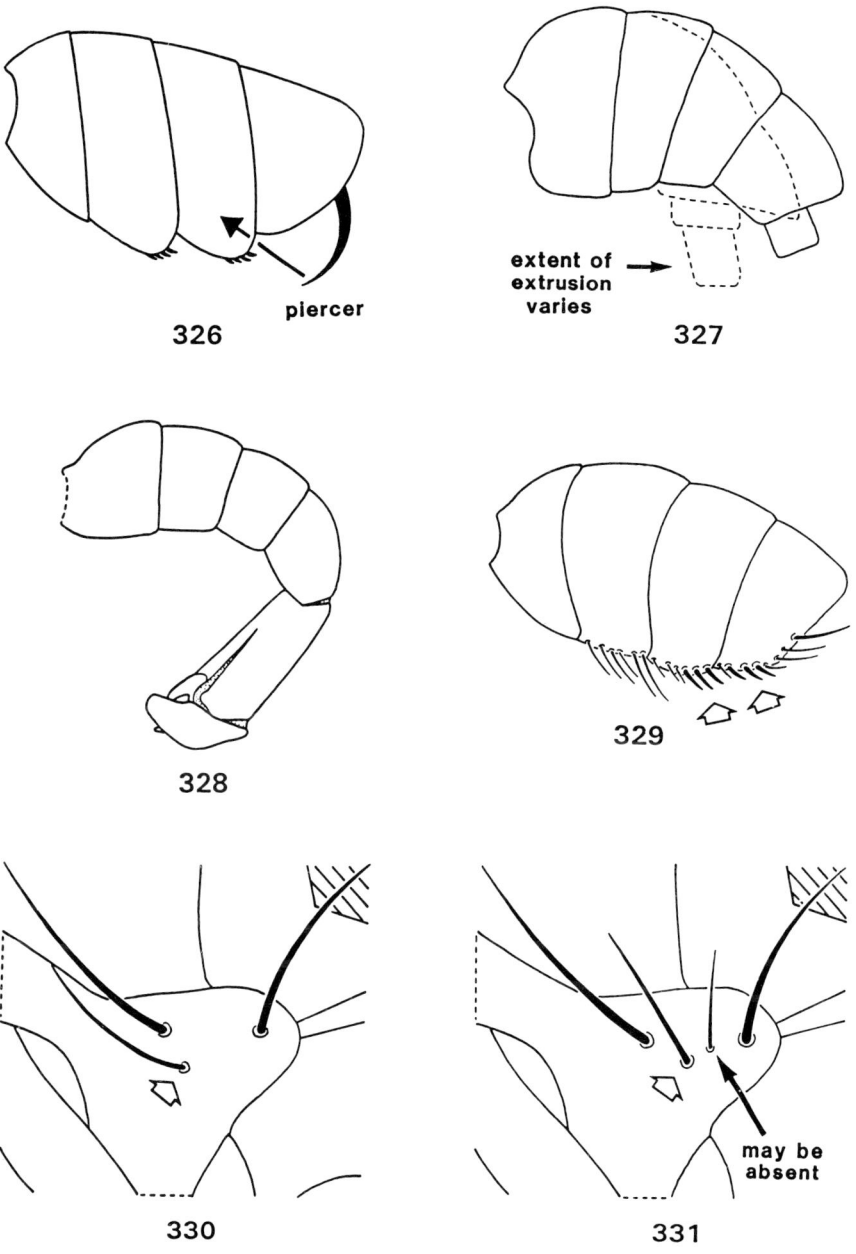

Figs 326–329. Female abdomen seen from the side.
326. *Compsilura concinnata*.
327. *Rondania fasciata* (dotted line shows the effect of shrinkage in dried specimens).
328. *Phania funesta*.
329. *Vibrissina debilitata*.
Figs 330–331. Bristles on katepisternum (see fig. 312 for orientation).
330. *Erycilla ferruginea*
331. *Erycia furibunda*.

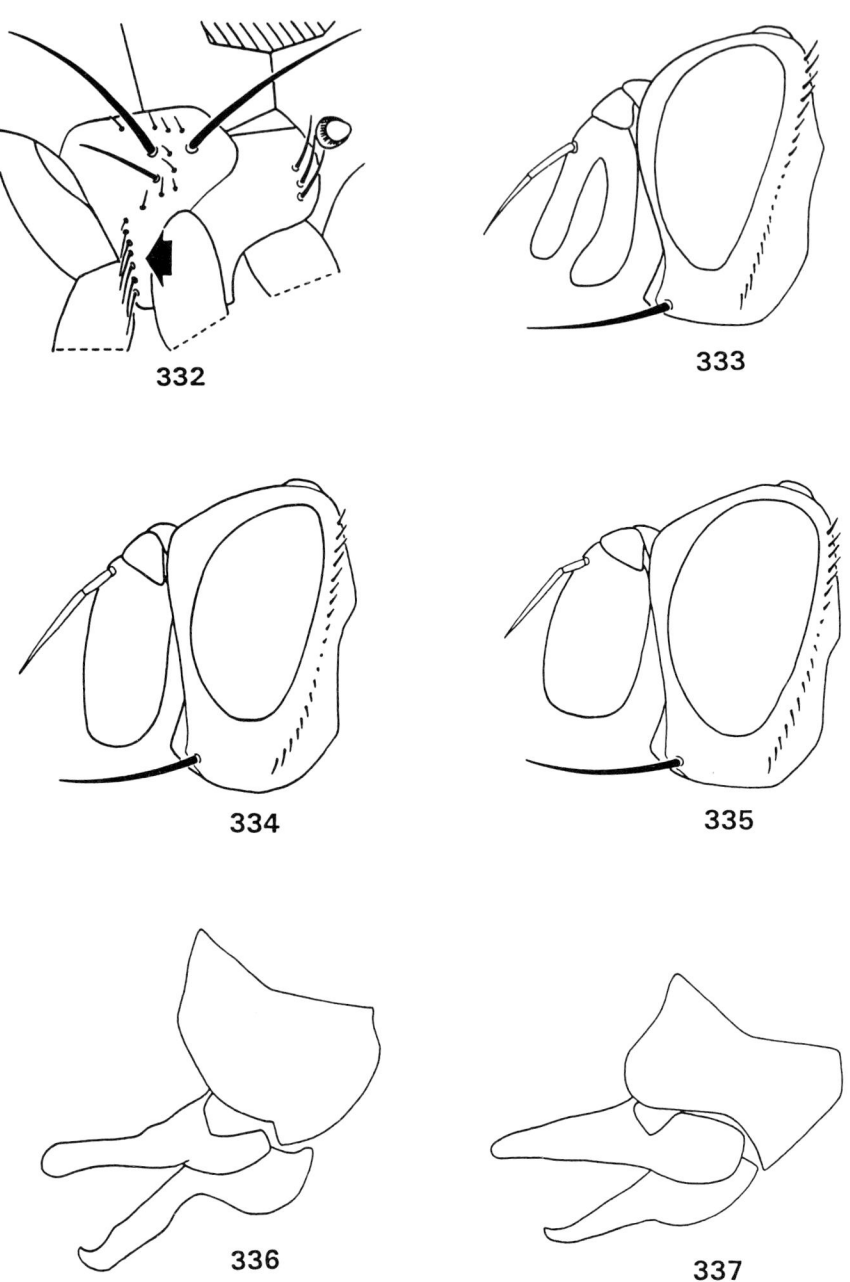

Fig. 332. *Actia lamia* katepisternum.
Fig. 334. *Actia crassicornis* male head.
Figs 336–337. Male terminalia (see fig.310 for orientation).
336. *Actia crassicornis.*

Fig. 333. *Peribaea fissicornis* male head.
Fig. 335. *Actia pilipennis* male head.

337. *Actia pilipennis.*

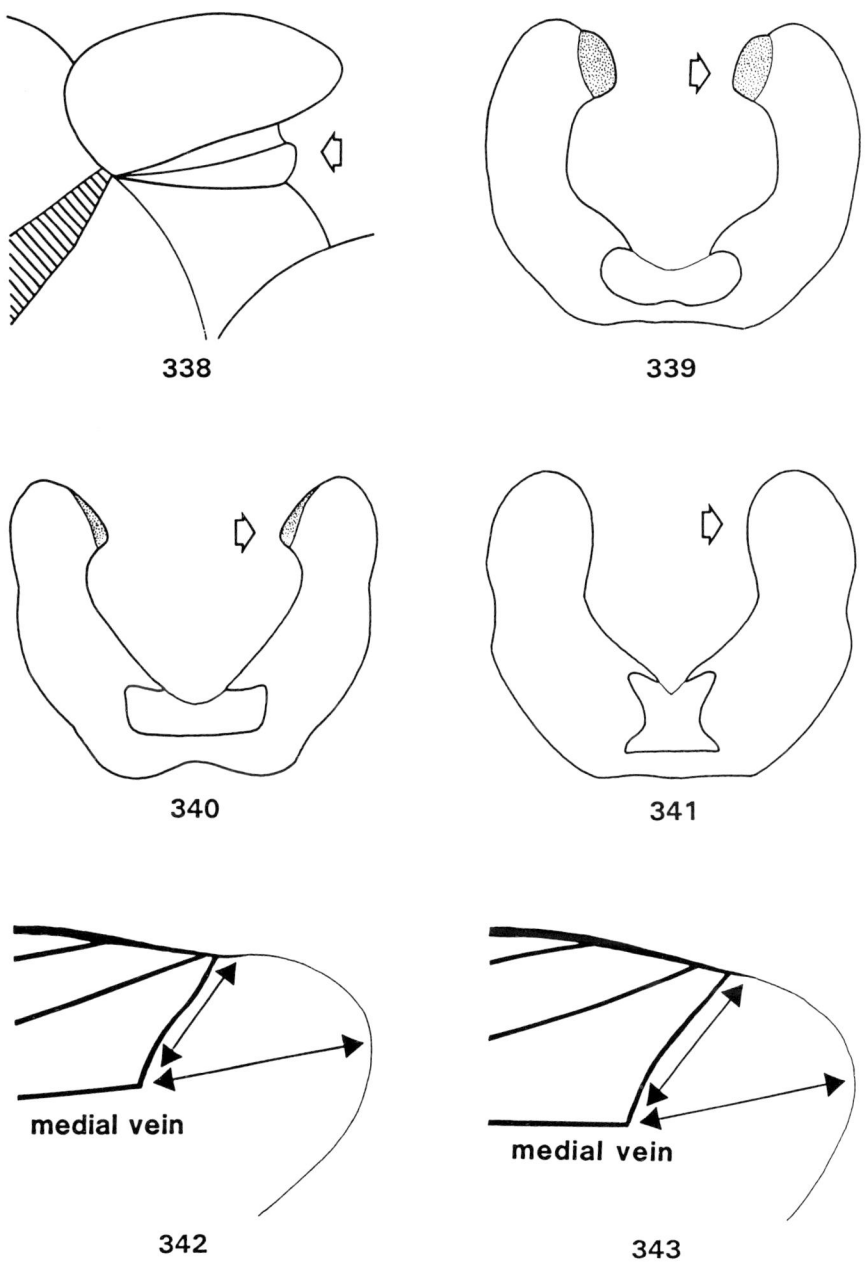

338

339

340

341

342

343

Fig. 338. *Phyto discrepans* (Rhinophoridae) subscutellum.
Figs 339–341. Male sternite 5 seen from below (see fig. 310 for orientation).

339. *Athrycia curvinervis*

340. *Athrycia impressa.*

341. *Athrycia trepida.*

Fig.342. *Athrycia impressa* wing.

Fig. 343. *Athrycia trepida* wing

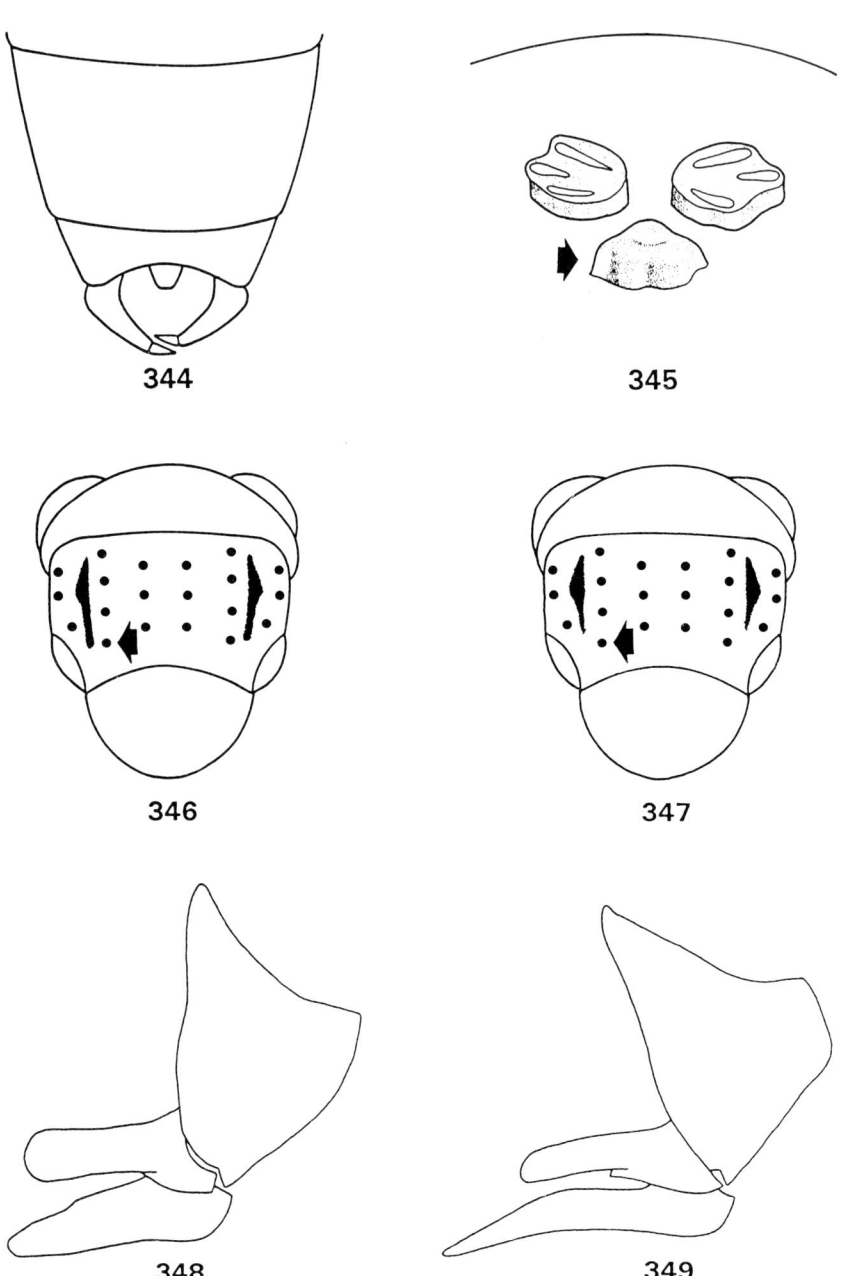

Fig. 344 *Labigastera forcipata* tip of female abdomen (seen from above).
Fig. 345. *Senometopia intermedia* puparium (area around posterior spiracles).
Figs. 346–347. Dorsal surface of thorax (viewed slightly from behind).
346. *Senometopia intermedia.*   347. *Senometopia pollinosa.*
Figs. 348–349. Male terminalia (see fig. 310 for orientation).
348. *Senometopia excisa.*   349. *Senometopia pollinosa*

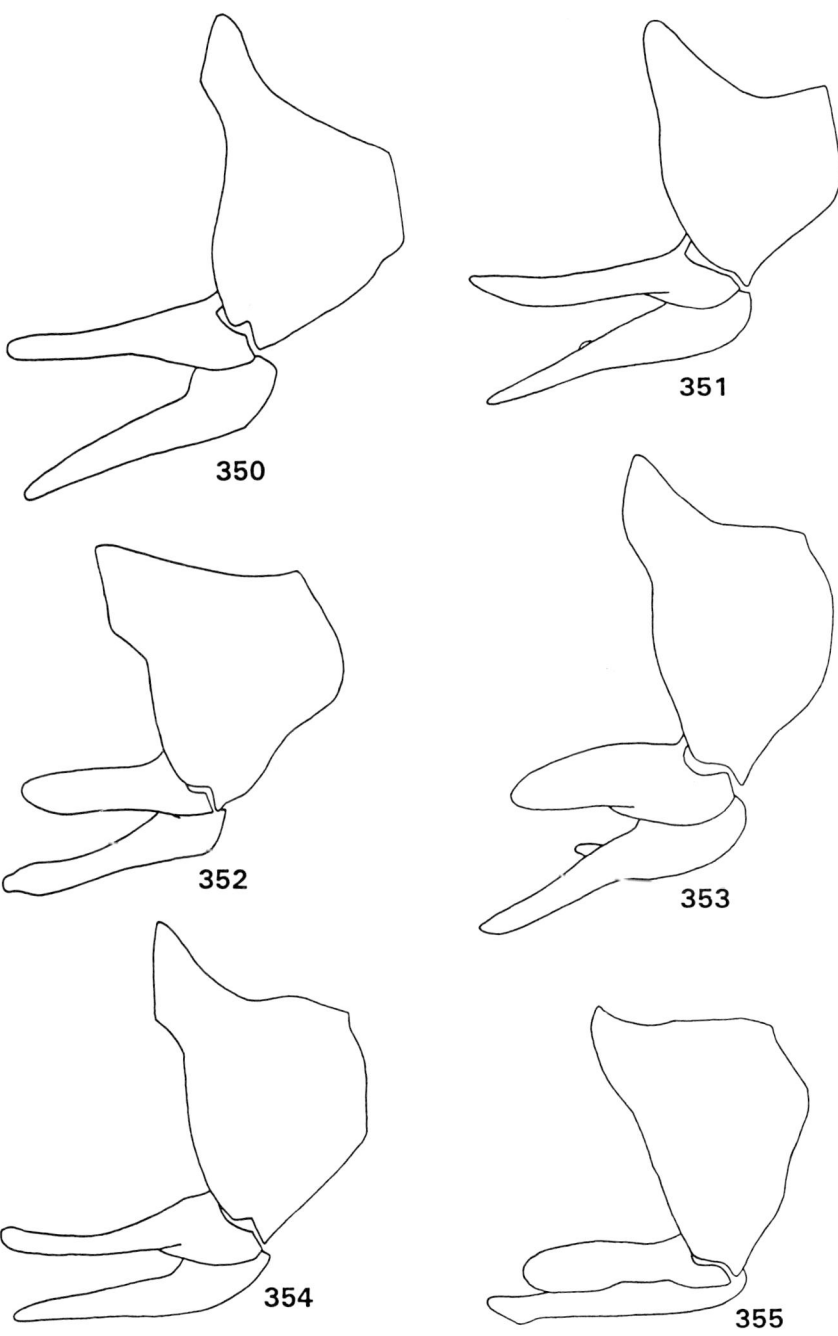

Figs. 350–355. Male terminalia (see fig. 310 for orientation).
350. *Carcelia rasa*.
352. *Carcelia tibialis*.
354. *Carcelia atricosta*.
351. *Carcelia puberula*.
353. *Carcelia lucorum*.
355. *Carcelia gnava*.

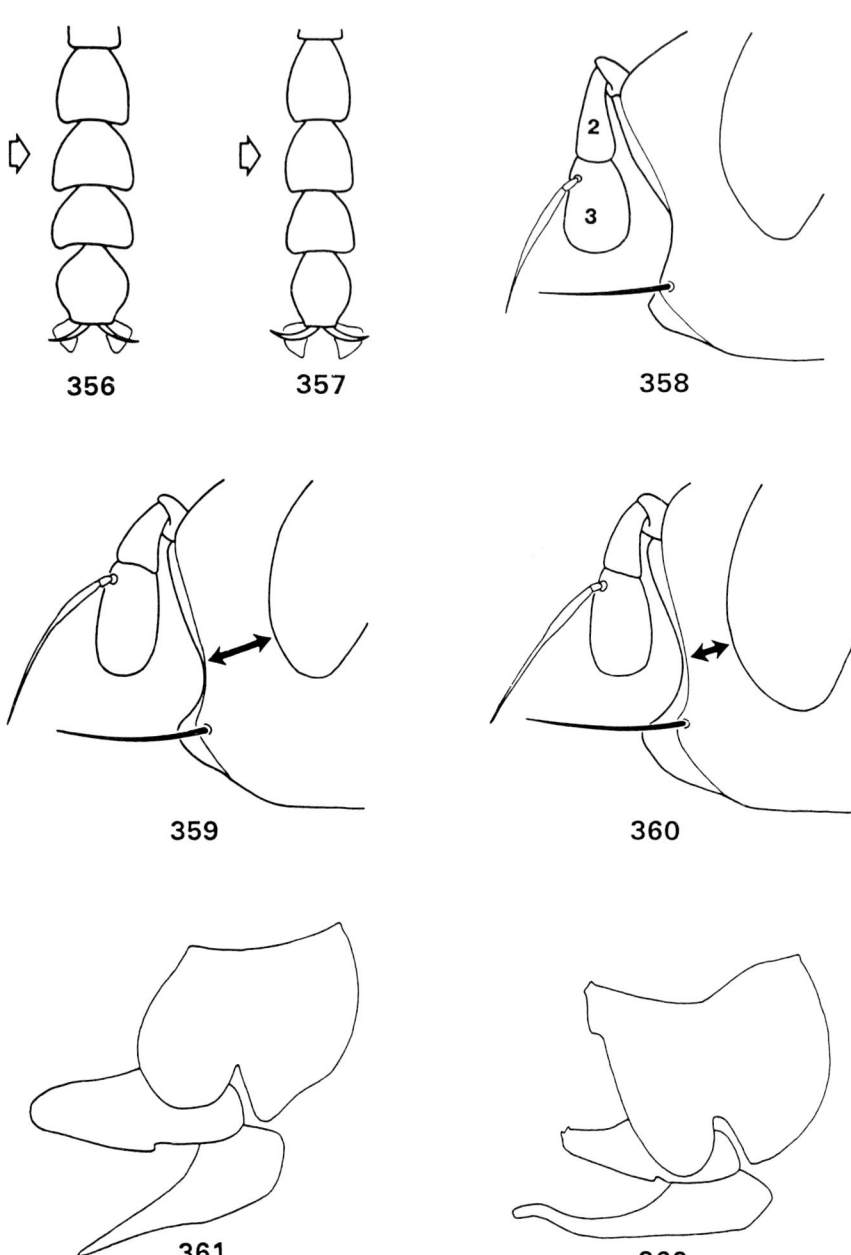

Figs 356–357. Tarsus of female front leg (seen from above).
356. *Ernestia rudis.*   357. *Ernestia laevigata.*
Figs 358–360. Head seen from the side.   358. *Eurithia caesia.*
359. *Eurithia intermedia* (female).   360. *Eurithia connivens* (female).
Figs 361–362. Male terminalia (see fig. 310 for orientation).
361. *Ernestia rudis.*   362. *Ernestia vagans.*

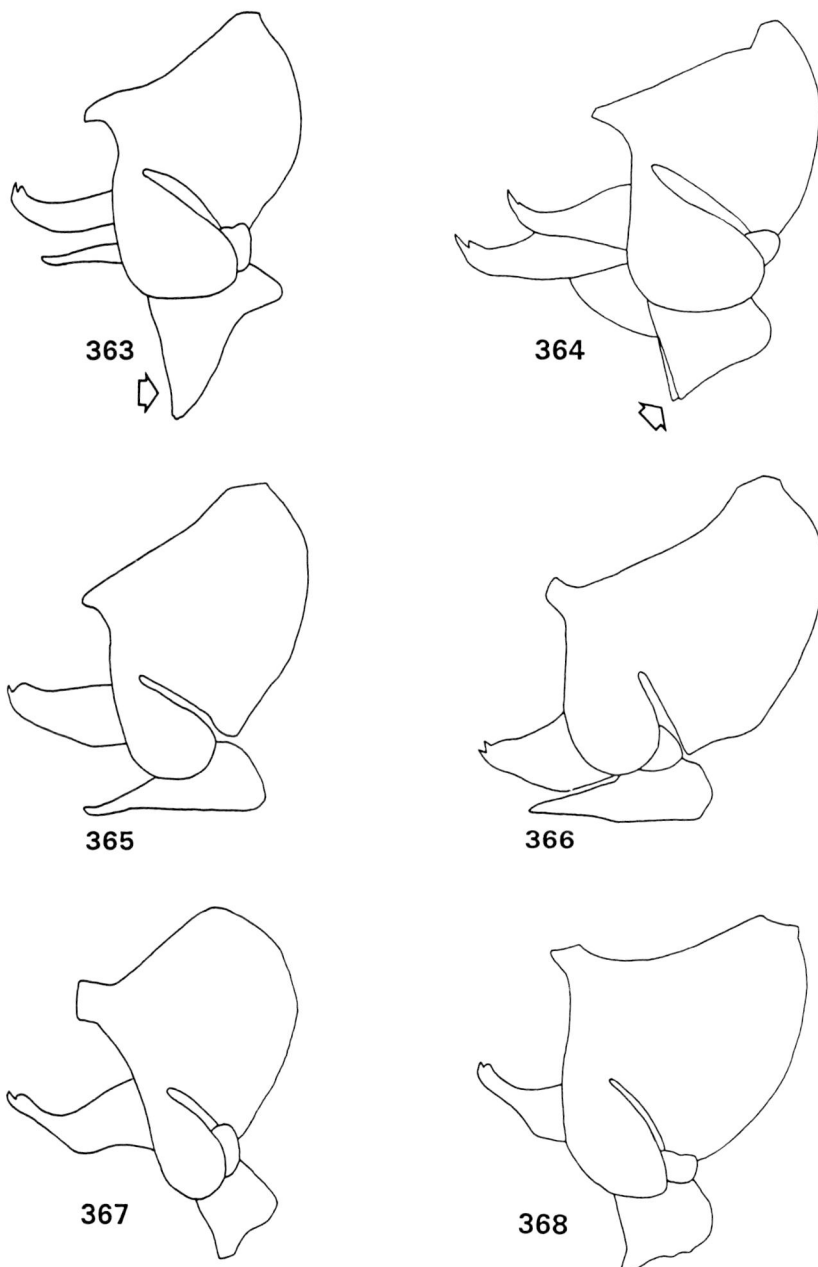

Figs 363–368. Male terminalia (see fig. 310 for orientation).
363. *Eurithia consobrina*.
364. *Eurithia vivida* (not seen directly from the side).
365. *Eurithia anthophila*.
366. *Eurithia caesia*.
367. *Eurithia intermedia*.
368. *Eurithia connivens*.

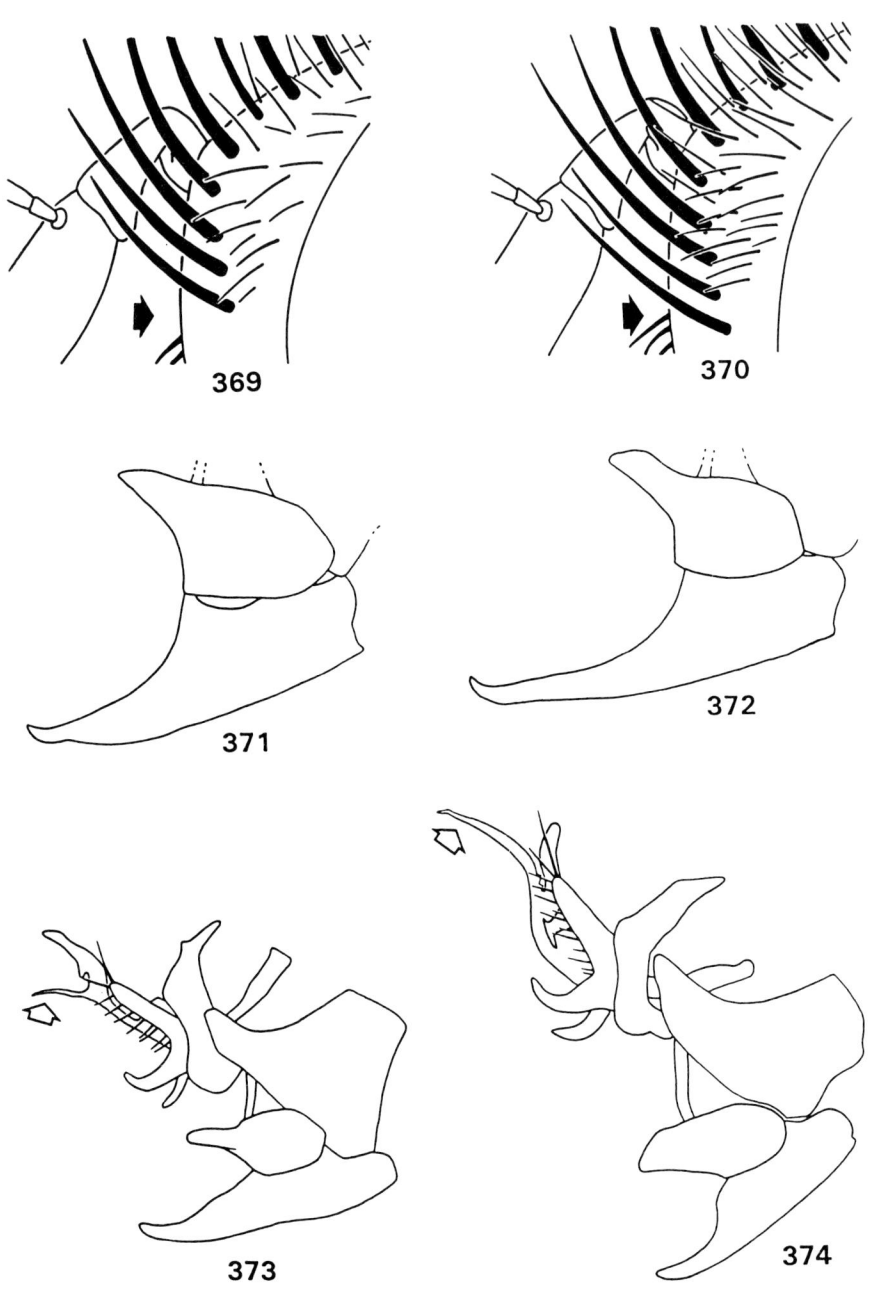

Figs 369–370. Male head seen from the side (see fig. 311 for orientation).
369. *Exorista larvarum.*   370. *Exorista fasciata.*
Figs 371–374. Male terminalia (see fig. 310 for orientation).
371. *Exorista larvarum.*   372. *Exorista fasciata.*
373. *Exorista glossatorum.*   374. *Exorista rustica*

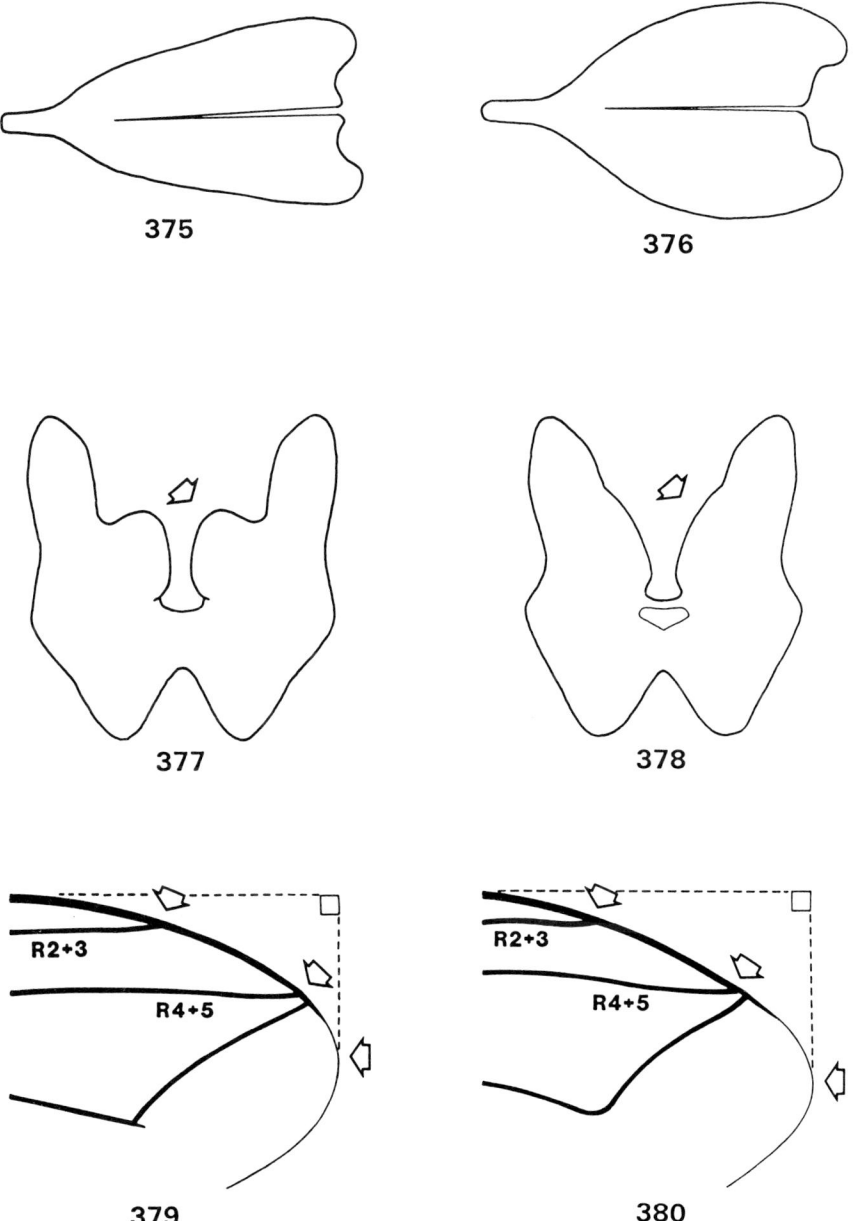

Figs 375–376. Male cerci seen from below (see fig. 310 for orientation).
375. *Exorista tubulosa.*   376. *Exorista rustica.*
Figs 377–378. Male sternite 5 seen from below (see fig. 310 for orientation).
377. *Exorista mimula.*   378. *Exorista rustica.*
Figs 379–380. Wing tip (upper dotted line is a continuation of the leading edge).
379. *Eumea linearicornis.*   380. *Platymya fimbriata.*

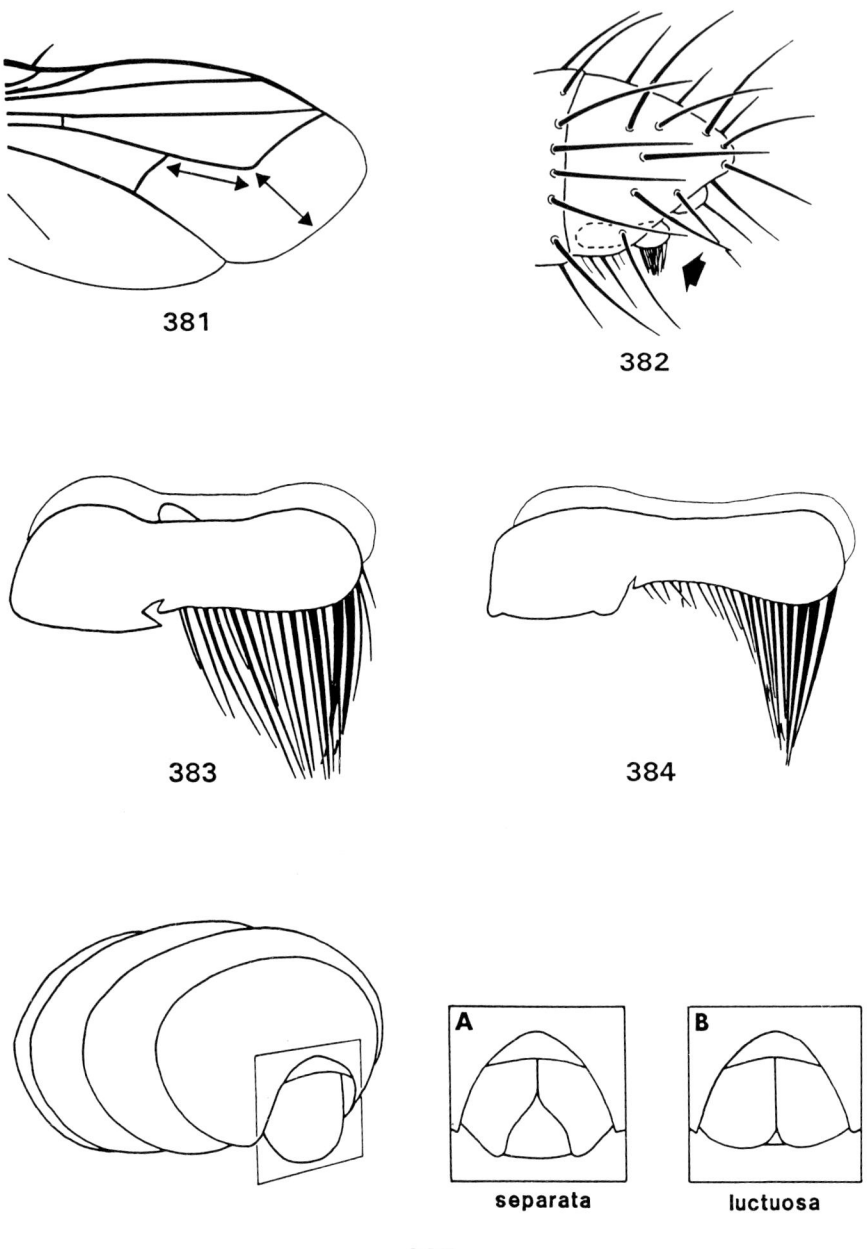

Fig. 381. *Brachicheta strigata* wing.

Fig. 382. *Medina* male abdomen (seen from the side with sternite 5 indicated).

Figs 383–384. Male sternite 5 seen from the side (see figs 382 and 310 for orientation). 383. *Medina luctuosa.* 384. *Medina separata.*

Fig. 385. *Medina* female abdomen (seen from behind); (a) *M.separata*; (b) *M.luctuosa.*

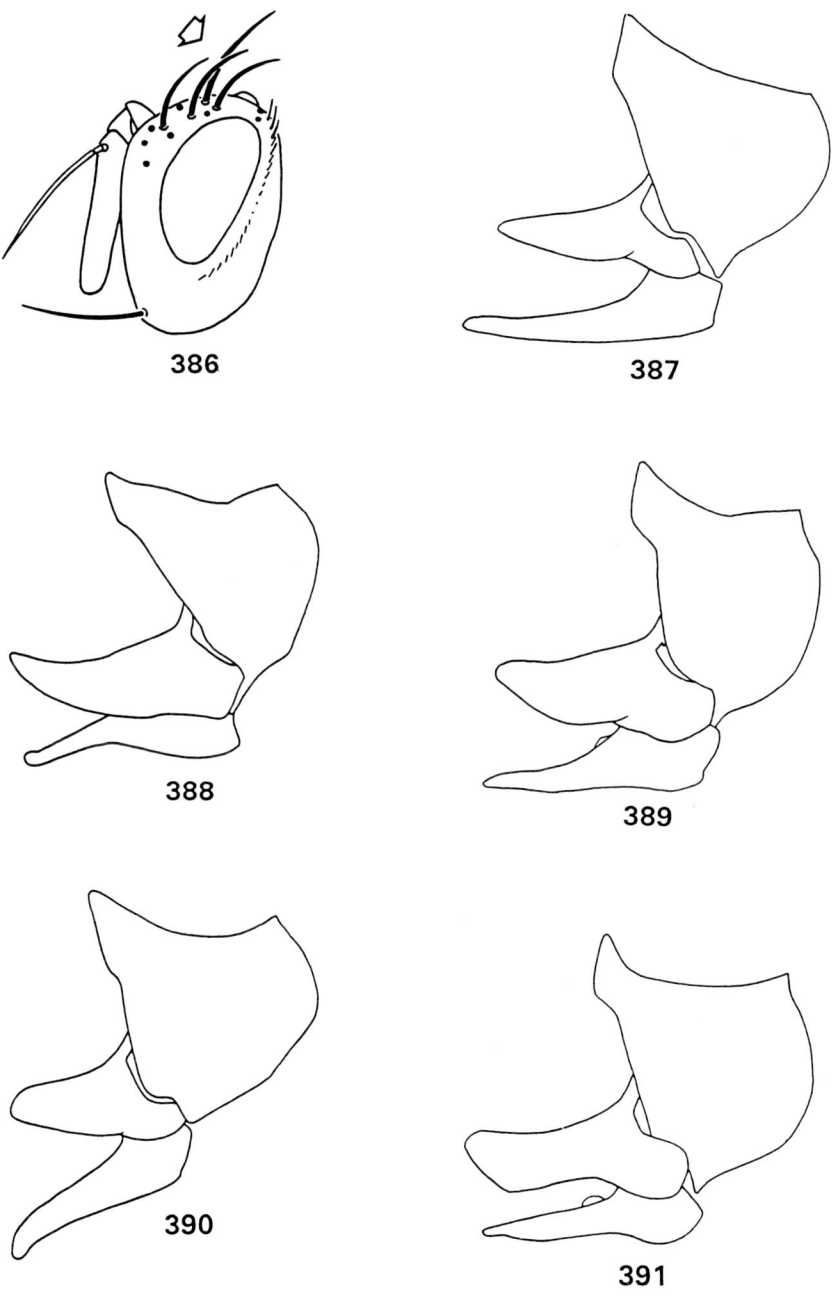

Fig. 386. *Thelymorpha marmorata* head (seen from the side).
Figs 387–391. Male terminalia (see fig. 310 for orientation, *vicina* after Wainwright, 1940).

387. *Phebellia glirina.*
388. *Phebellia vicina.*
389. *Phebellia villica.*
390. *Phebellia glauca.*
391. *Phebellia stulta.*

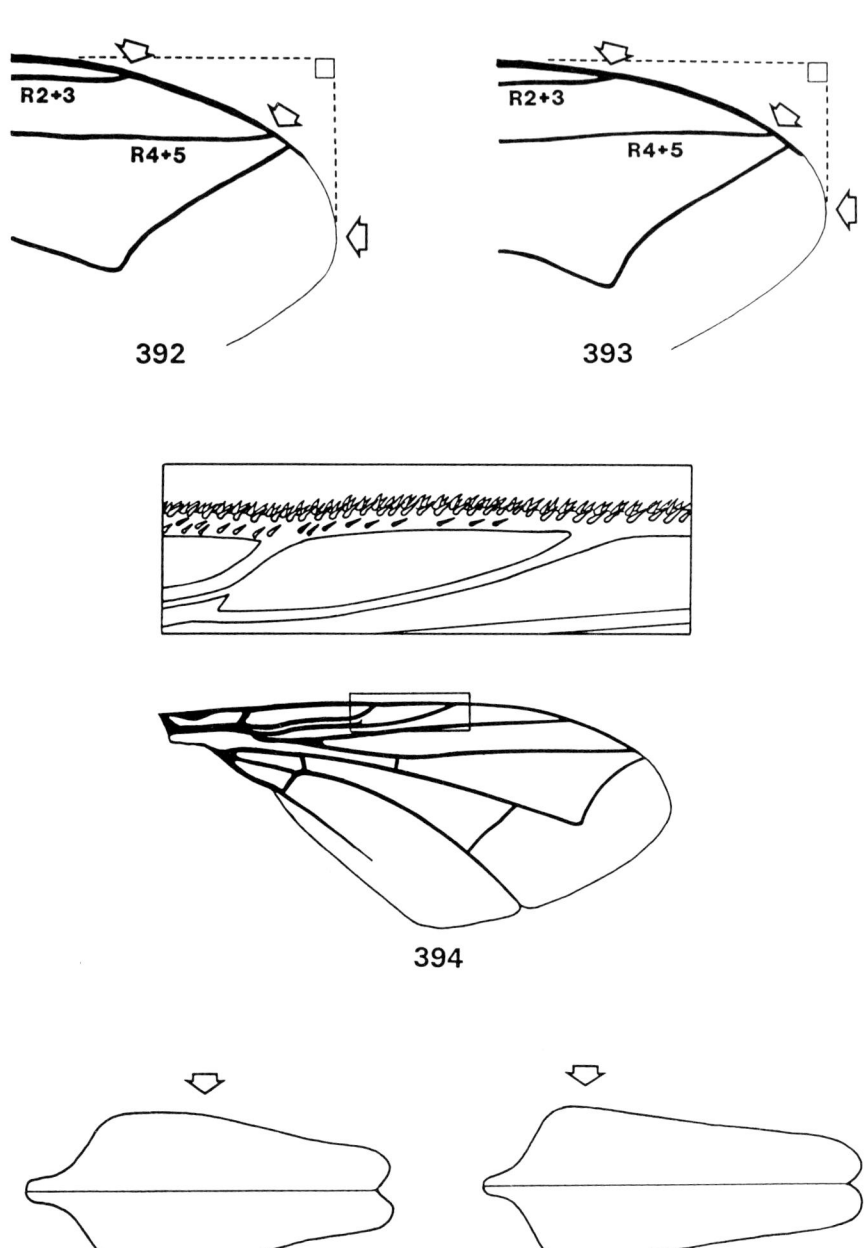

Figs 392–393. Wing tip (upper dotted line is a continuation of the leading edge).
392. *Phebellia glirina.*   393. *Phebellia glauca.*
Fig. 394. *Phryxe heraclei* wing (expansion shows underside of subcostal vein).
Figs 395–396. Male cerci seen from below (see fig. 310 for orientation).
395. *Phorocera assimilis.*   396. *Phorocera obscura.*

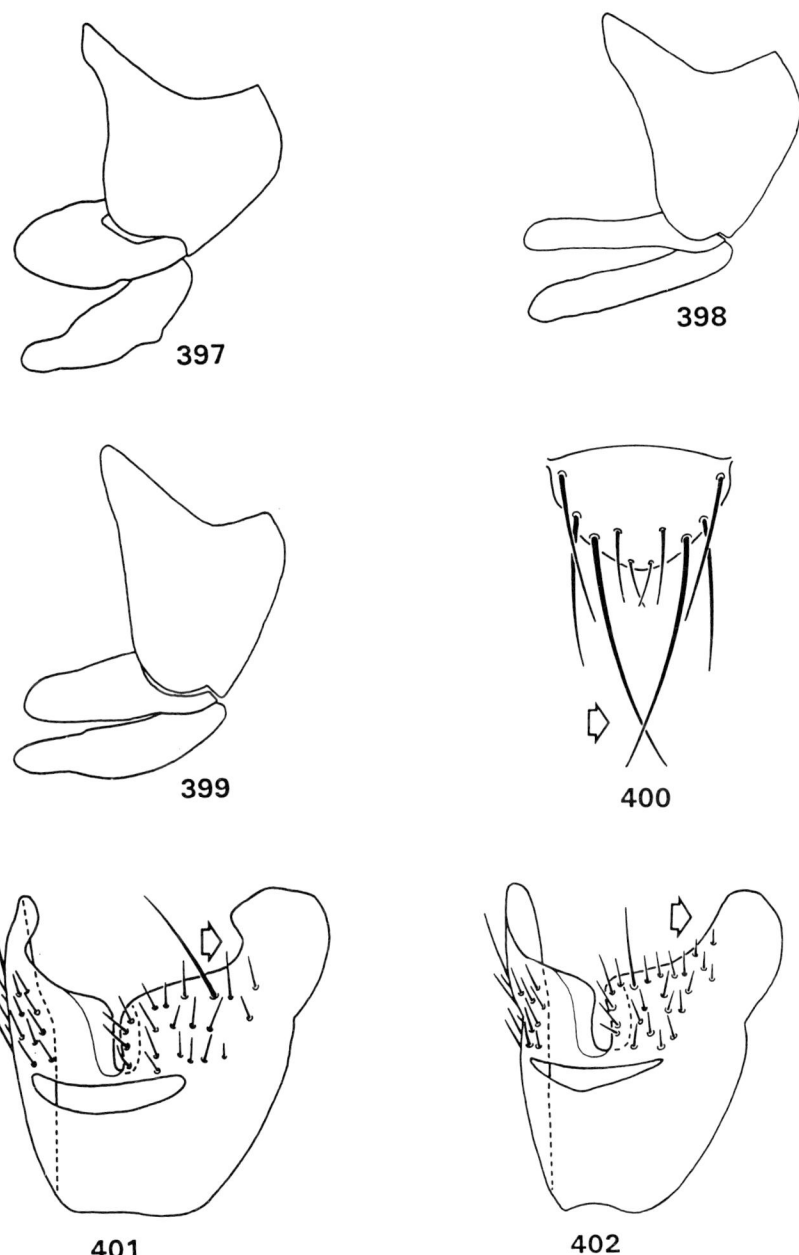

Figs 397–399. Male terminalia (see fig. 310 for orientation).
397. *Phryxe heraclei.*   398. *Phryxe magnicornis.*
399. *Phryxe vulgaris.*
Fig. 400. Generalised Siphonini scutellum (longest pair of bristles not always crossed).
Figs 401–402. Male sternite 5 seen from below and slightly to one side (see fig. 310).
401. *Siphona mesnili.*   402. *Siphona cristata.*

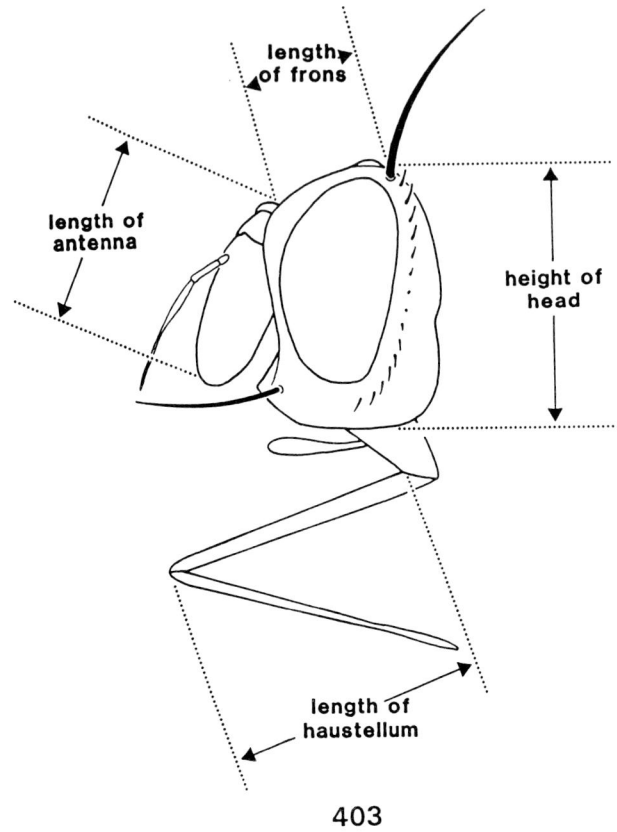

403

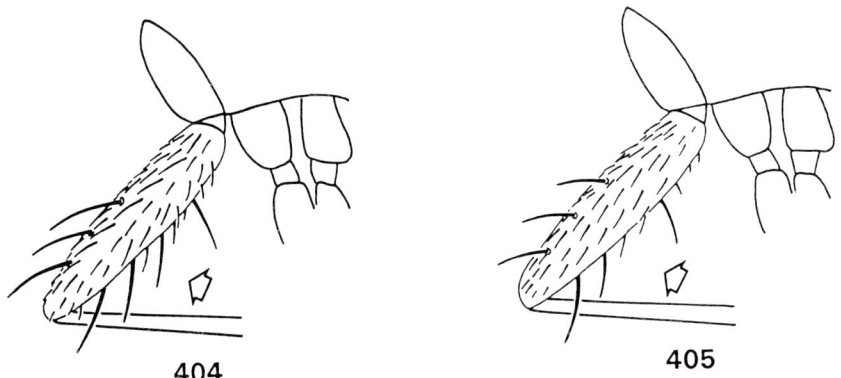

404

405

Fig. **403.** *Siphona* head seen from the side.
Figs **404–405.** Femur of left-side front leg (seen from behind).
404. *Siphona mesnili.*   405. *Siphona cristata*

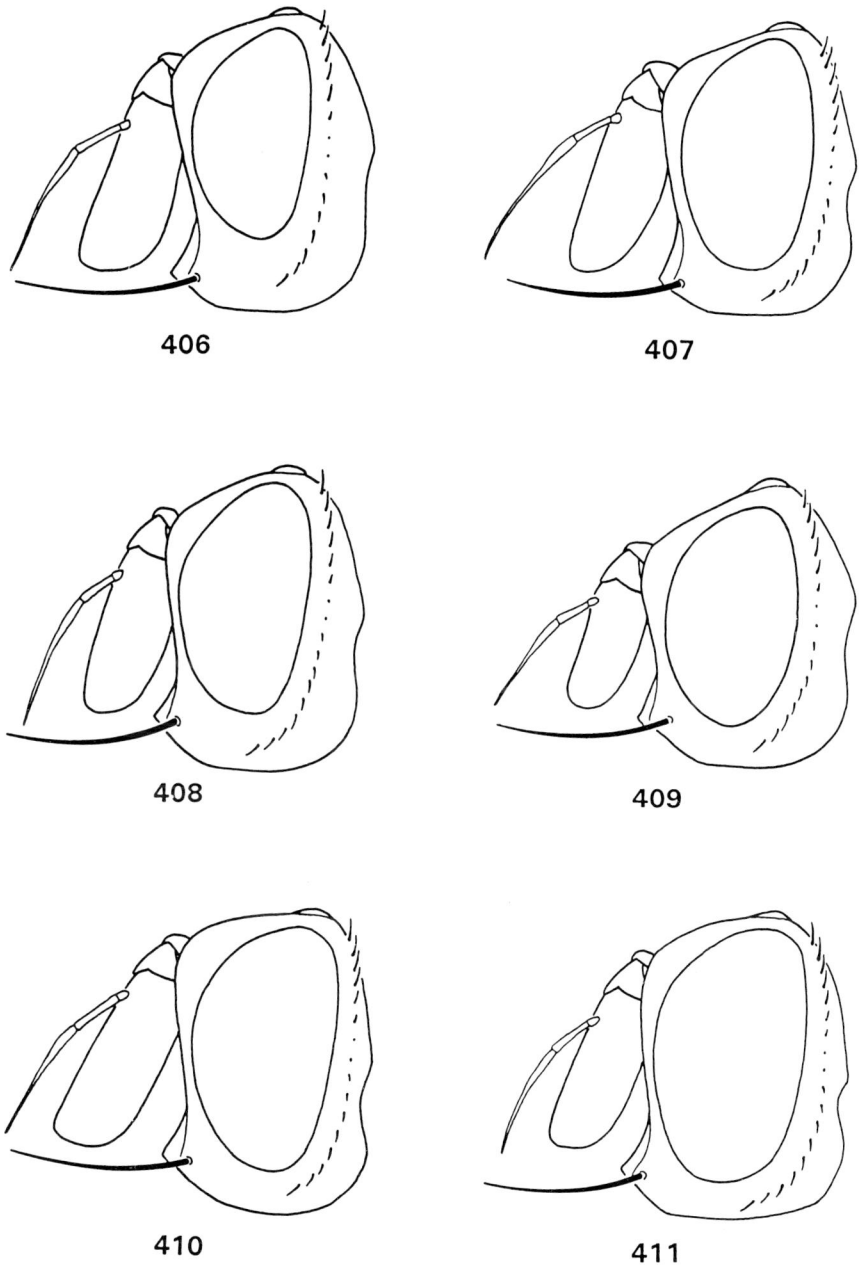

Figs 406–411. Head seen from the side.
406. *Siphona ingerae.*
408. *Siphona mesnili.*
410. *Siphona setosa.*

407. *Siphona maculata.*
409. *Siphona geniculata.*
411. *Siphona pauciseta.*

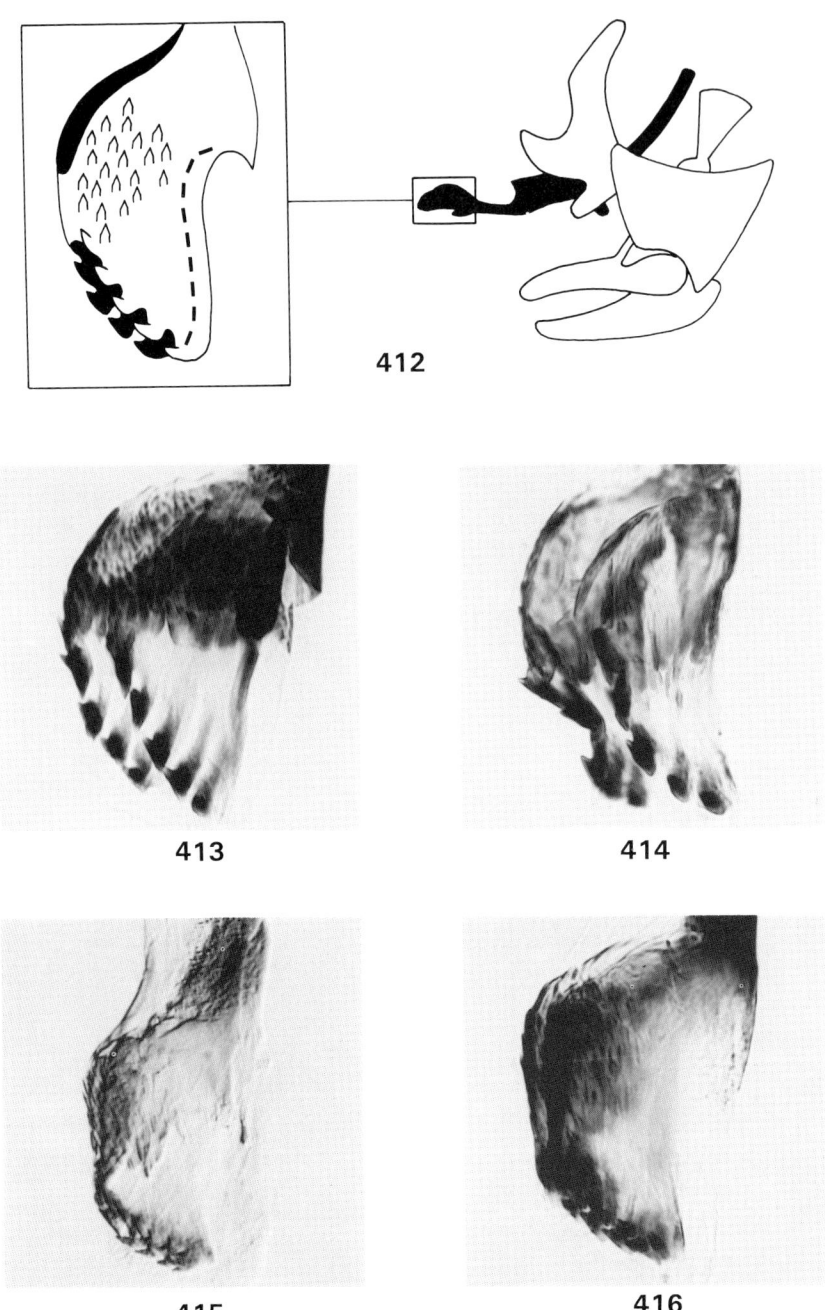

412

413

414

415

416

Fig. 412. Generalised male terminalia of *Siphona*.
Figs 413–416. Distiphalli seen from the side (see fig. 412 for orientation).
413. *Siphona ingerae*.   414. *Siphona maculata*.
415. *Siphona collini*.   416. *Siphona mesnili*.

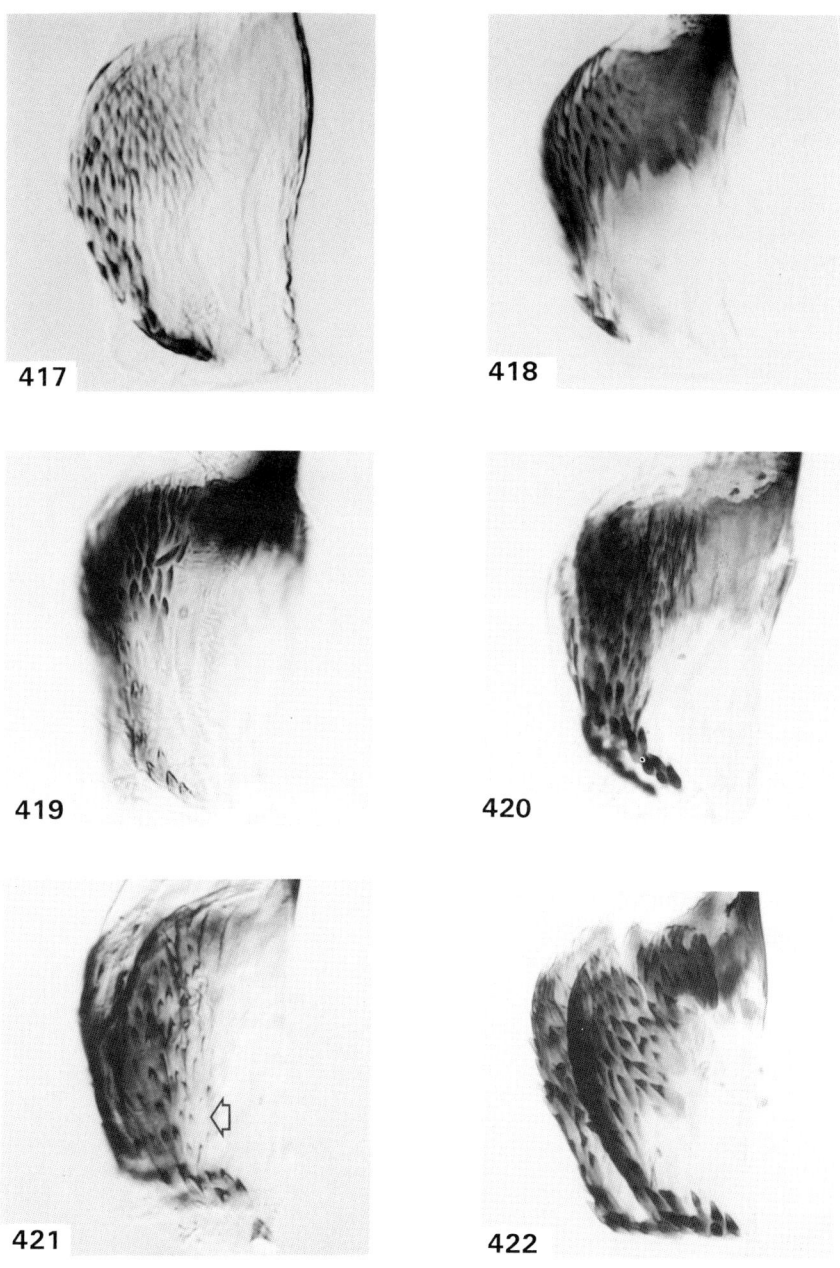

Figs 417–422. Distiphalli seen from the side (see fig. 412 for orientation).
417. *Siphona cristata.*
418. *Siphona variata.*
419. *Siphona geniculata.*
420. *Siphona setosa**.
421. *Siphona pauciseta.*
422. *Siphona boreata**.
* apical row of teeth may be single or (as shown here) double.

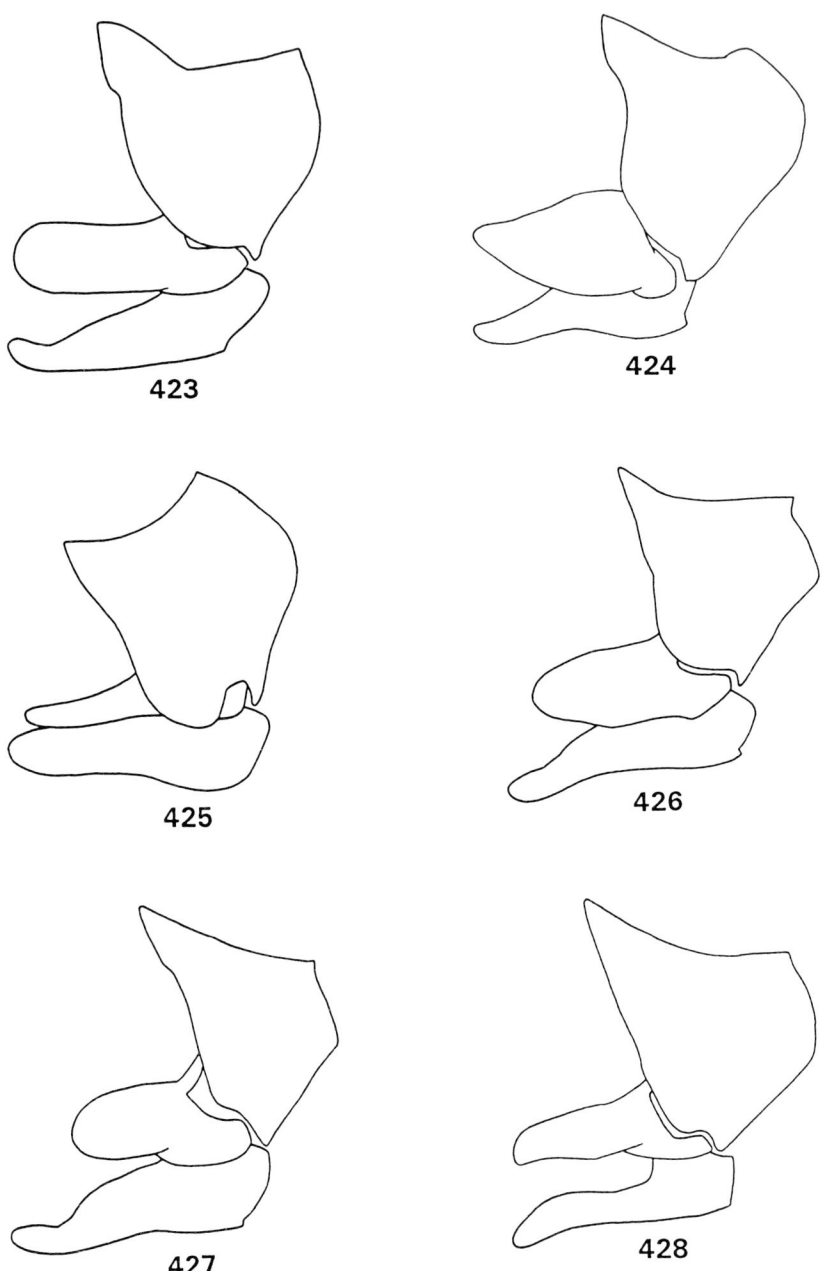

Figs 423–428. Male terminalia (see fig. 310 for orientation).
423. *Pseudoperichaeta nigrolineata.*   424. *Clemelis pullata.*
425. *Nilea hortulana.*   426. *Eumea linearicornis.*
427. *Platymya fimbriata.*   428. *Aplomya confinis.*

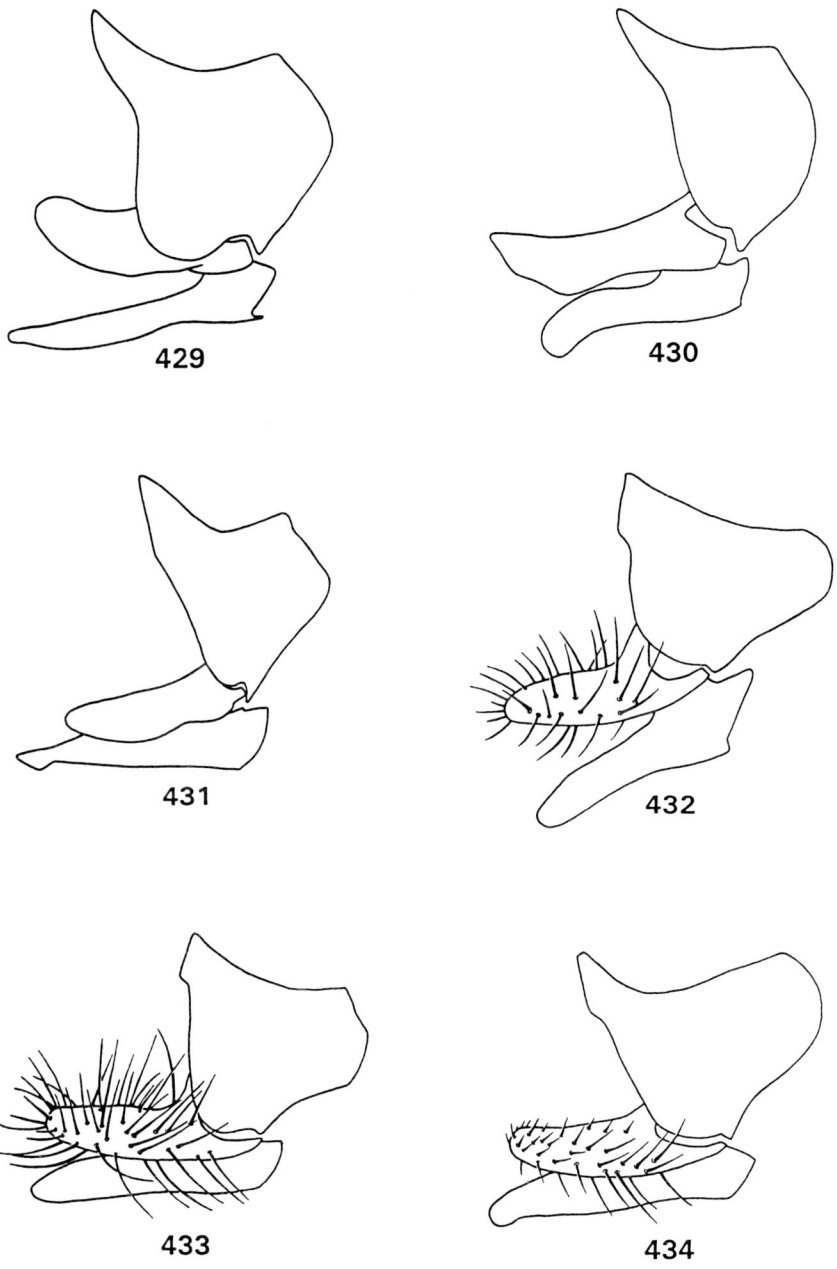

Figs 429–434. Male terminalia (see fig. 310 for orientation).
429. *Tlephusa cincinna.*
431. *Huebneria affinis.*
433. *Meigenia dorsalis.*
430. *Zenillia libatrix.*
432. *Meigenia majuscula.*
434. *Meigenia mutabilis.*

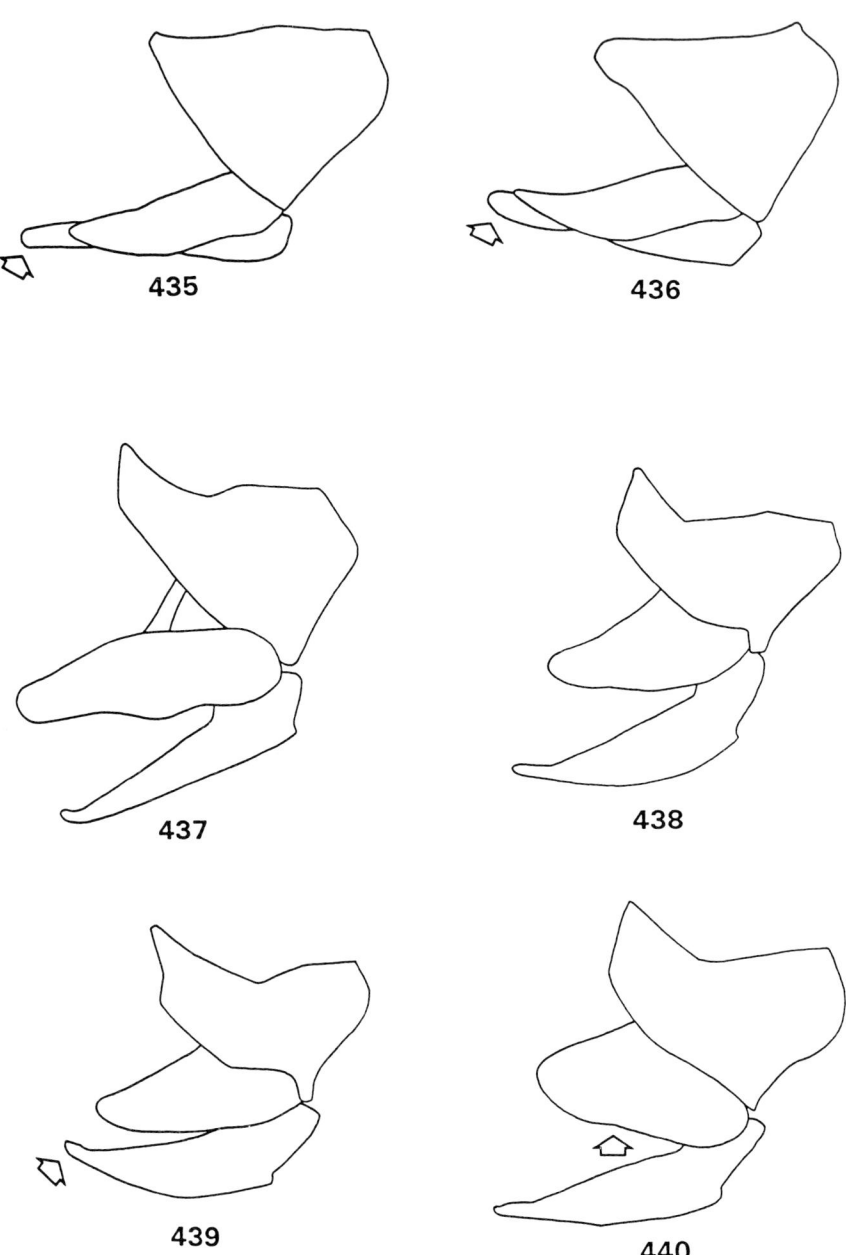

Figs 435–440. Male terminalia (see fig. 310 for orientation).
435. *Myxexoristops stolida.*
436. *Myxexoristops blondeli*
437. *Timavia amoena.*
438. *Winthemia variegata.*
439. *Winthemia quadripustulata.*
440. *Winthemia cruentata.*

References

ALDRICH, J.R., KOCHANSKY, J.P. & ABRAMS, C.B. 1984. Attractant for a beneficial insect and its parasitoids: pheromone of the predatory Spined Soldier Bug, *Podisus maculiventris* (Hemiptera: Pentatomidae). *Environmental Entomology* **13**: 1031–1036.

ALLEN, A.A. 1963. *Alophora obesa* F. (Dipt., Tachinidae) bred from *Neottiglossa pusilla* Gmel. (Hem., Pentatomidae). *Entomologist's Monthly Magazine* **99**: 35.

ALLEN, A.A. 1966. *Subclytia rotundiventris* Fall. (Dipt., Tachinidae): a new host and county record. *Entomologist's Monthly Magazine* **102**: 198.

ALLEN, A.A. 1987. *Lophosia fasciata* Mg. (Dipt.: Tachinidae) in the London suburbs, and an apparently new host record. *Entomologist's Record and Journal of Variation* **99**: 83.

ALMA, P.J. 1976. Parasitization of *Tipula* spp. (Dipt., Tipulidae) by *Siphona geniculata* (DeGeer) (Dipt., Tachinidae). *Entomologist's Monthly Magazine* **111**: 105–107.

ANDERSEN, S. 1982. Revision of European species of *Siphona* Meigen (Diptera: Tachinidae). *Entomologica Scandinavica* **13**: 149–172.

ANDERSEN, S. 1983. Phylogeny and classification of Old World genera of Siphonini (Diptera: Tachinidae). *Entomologica Scandinavica* **14**: 1–15.

ANDERSEN, S. 1988. Revision of European species of *Phytomyptera* Rondani (Diptera; Tachinidae). *Entomologica Scandinavica* **19**: 43–80.

ARNAUD, P.H., JNR. 1963. *Perumyia embiaphaga*, a new genus and species of Neotropical Tachinidae (Diptera) parasitic on Embioptera. *American Museum Novitates* **2143**: 1–9.

ARNAUD, P.H., JNR. 1978. A host-parasite catalog of North American Tachinidae (Diptera). *Miscellaneous Publications. United States Department of Agriculture* **1319**: 1–860.

AUDCENT, H. 1932. Bristol insect fauna: Diptera (part 5). *Proceedings of the Bristol Naturalists' Society* **7**: 358–370.

AUDCENT, H. 1942. A preliminary list of the hosts of some British Tachinidae (Dipt.). *Transactions of the Society for British Entomology* **8**: 1–42.

BANKS, C.J. 1956. A second record of a tachinid (Dipt.) parasite bred from one of the Coccinellinae (Col., Coccinellidae). *Entomologist's Monthly Magazine* **92**: 188.

BARFOOT, S.D. 1957. Oviposition of *Ptychomyia selecta* Meig. (Dipt., Tachinidae). *Entomologist's Monthly Magazine* **93**: 95

BAYNE, D.M. 1987. *Alophora hemiptera* (F.) (Dipt., Tachinidae) in Scotland. *Entomologist's Monthly Magazine* **123**: 81.

BELSHAW, R. 1992. Tachinid (Diptera) assemblages in habitats of a secondary succession in southern Britain. *Entomologist* **111**: 151–161.

BERRY, P.A. & PARKER, H.L. 1950. Notes on parasites of *Sitona* in Europe, with especial reference to *Campogaster exigua* (Meig.). *Proceedings of the Entomological Society of Washington* **52**: 251–258.

BRADLEY, J.D. & FLETCHER, D.S. 1986. *An indexed list of British butterflies and moths*. Orpington: Kedleston Press.

BRADLEY, J.D., TREMEWAN, W. G. & SMITH, A. 1979. *British Tortricoid Moths. Tortricidae: Olethreutinae*. London: The Ray Society.

BROWNING, H.W. & OATMAN, E.R. 1984. Intra- and interspecific relationships among some parasites of *Trichoplusia ni* (Lepidoptera: Noctuidae). *Environmental Entomology* **13**: 551–556.

CADE, W.H. 1975. Acoustically orientating parasitoids: fly phonotaxis to cricket song. *Science* **190**: 1312–1313.

CAMERON, A.E., McHARDY, J.W., & BENNET, A.H. 1944. *The Heather Beetle (Lochmaea suturalis)*. Petworth, Sussex: British Field Sports Society.

CANTRELL, B.K. 1988. The comparative morphology of the male and female postabdomen of the Australian Tachinidae (Diptera), with descriptions of some first-instar larvae and pupae. *Invertebrate Taxonomy* **2**: 81–221.

CANTRELL, B.K. & CROSSKEY, R.W. 1989. Family Tachinidae. pp 733–784 *In* Evenhuis N.L. [Ed.]. *Catalogue of the Diptera of the Australasian and Oceanean regions*. Honolulu and Leiden: Bishop Museum Press and E.J. Brill.

CARR, J.W. 1935. *The invertebrate fauna of Nottinghamshire: supplement*. Nottingham: J. and H. Bell.

CHANDLER, P.J. 1966. Notes on some uncommon Diptera. *Entomologist's Record and Journal of Variation* **78:** 80.

CHANDLER, P. 1976. Notes on some uncommon Calyptrate flies (Diptera) observed during recent years. *Entomologist's Record and Journal of Variation* **88:** 14–19.

CHENG, L. 1967. Notes on three species of *Actia* (Dipt., Tachinidae), parasites of oak-feeding caterpillars. *Entomologist* **100:** 265–268.

CHENG, L. 1969. The life history and development of *Lypha dubia* Fall. (Dipt.,Tachinidae). *Entomologist* **101:** 25–32.

CHENG, L. 1970. Timing of attack by *Lypha dubia* Fall. (Diptera: Tachinidae) on the Winter Moth *Operophtera brumata* (L.) (Lepidoptera: Geometridae) as a factor affecting parasite success. *Journal of Animal Ecology* **39:** 313–320.

CHISWELL, J.R. 1956. A taxonomic account of the final instar larvae of some British Tipulinae (Diptera; Tipulidae). *Transactions of the Royal Entomological Society of London* **108:** 409–484.

CLAUSEN, C.P. 1940. *Entomophagous insects.* New York and London: McGraw-Hill.

CLEMENT, S.L., RUBINK, W.L. & McCARTNEY, D.A. 1986. Larviposition response of *Bonnetia comta* (Dipt.: Tachinidae) to a kairomone of *Agrotis ipsilon* (Lep.: Noctuidae). *Entomophaga* **31:** 277–284.

CLEMONS, L. 1992. *Litophasia hyalipennis* (Fallen) (Diptera: Tachinidae) in North Kent. *Entomologist's Record and Journal of Variation* **104:** 201-202.

COLLIN, J.E. 1945. Parasitism of a *Merodon* larva by the tachinid *Lypha dubia* Fln. *Entomologist's Record and Journal of Variation* **57:** 70.

CROSSKEY, R.W. 1976. A taxonomic conspectus of the Tachinidae (Diptera) of the Oriental region. *Bulletin of the British Museum (Natural History). Entomology,* Supplement **26:** 1–357.

CRUTTWELL, R.E. 1969. The biology and mode of parasitism of *Uromacquartia trinitatis* Thompson (Diptera: Tachinidae). *Technical Bulletin. Commonwealth Institute of Biological Control* **12:** 20–28.

D'ASSIS-FONSECA. see FONSECA

DAVIES, A.J. 1986. Multiparasitism of *Pleuroptya ruralis* (Scopoli) (Lep., Pyralidae), by *Nemorilla floralis* Fall. and *Phryxe nemea* (Meig.) (Dipt., Tachinidae). *Entomologist's Monthly Magazine* **122:** 72.

DAY, C.D. 1948. *British tachinid flies.* Arbroath: T.Buckle. [Also *North Western Naturalist* **21** and **22** (1946 and 1947).]

DeVRIES, P.J. 1984. Butterflies and Tachinidae: does the parasite always kill its host? *Journal of Natural History* **18:** 323–326.

DOWDEN, P.B. 1933. *Lydella nigripes* and *L.piniariae*, fly parasites of certain tree-defoliating caterpillars. *Journal of Agricultural Research* **46:** 963–995.

DOWDEN, P.B. 1934. *Zenillia libatrix* Panz., a tachinid parasite of the Gypsy Moth and the Brown Tail Moth. *Journal of Agricultural Research* **48:** 97–114.

DRUMMOND, D.C. 1952. *Macquartia tenebricosa ab. nitida* Meig. (Dipt., Tachinidae) bred from *Chrysolina gramminis* L. (Col., Chrysomelidae). *Entomologist's Monthly Magazine* **88:** 46.

DUPUIS, C. 1963. Essai monographique sur les Phasiinae (Diptères Tachinaires parasites d'Hétéroptères). *Mémoires. Muséum National d'Histoire Naturelle.* Nouvelle Série. (Série A, Zoologie) **26:** 1–461.

EGGLETON, P. & GASTON, K.J. 1992. Tachinid host ranges: a reappraisal (Diptera: Tachinidae). *Entomologist's Gazette* **43:** 139–143.

ELLIS 1926. *Viviana cinerea* Fallén, parasitic on *Carabus monilis* F. *Proceedings of the Entomological Society of London* **1:** 54–55.

ELSEY, K.D. & RABB, R.L. 1970. Biology of *Voria ruralis* (Diptera: Tachinidae). *Annals of the Entomological Society of America* **63:** 216-222.

EMDEN, F.I. VAN 1950. Dipterous parasites of Coleoptera. *Entomologist's Monthly Magazine* **86:** 182–206.

EMDEN, F.I. VAN 1954. Diptera Cyclorrhapha Calyptrata (1), section (a): Tachinidae and Calliphoridae. *Handbooks for the Identification of British Insects* **10** (4a). 1–133.

EMMET, A.M. [Ed.] 1988. *A field guide to the smaller British Lepidoptera.* London: The British Entomological and Natural History Society.

ENGLISH-LOEB, G.M., KARBAN, R. & BRODY, A.K. 1990. Arctiid larvae survive attack by a tachinid parasitoid and produce viable offspring. *Ecological Entomology* **15:** 361–362.

EYLES, A.C. 1962. Some notes on natural enemies of Lygaeidae. *Entomologist's Monthly Magazine* **98:** 226–227.

FALK, S. in press. *A review of the scarce and threatened flies of Great Britain* **2.** Peterborough: Nature Conservancy Council.

FERRAR, P. 1977. Parasitism of other adult Diptera by Tachinidae in Australia. *Journal of the Australian Entomological Society* **16:** 397–401.

FERRAR, P. 1987. *A guide to the breeding habits and immature stages of Diptera Cyclorrhapha*. Leiden and Copenhagen: E.J. Brill/ Scandinavian Science Press.

FONSECA, E.C.M. D'ASSIS– 1949. *Dionaea aurifons* Mg. (Dipt., Larvaevoridae) in North Devon. *Entomologist's Monthly Magazine* **85:** 19–20.

FORD, T.H. 1973. Some records of bred Tachinidae. *Entomologist's Record and Journal of Variation* **85:** 288–299.

FORD, T.H. 1976. Some records of bred Tachinidae − 2. *Entomologist's Record and Journal of Variation* **88:** 68–71.

FORD, T.H. 1989. *Medina separata* (Meigen) (Dipt., Tachinidae), new to Britain. *Entomologist's Monthly Magazine* **125:** 139–140.

FORD, T.H. & SHAW, M.R. 1991. Host records of some West Palaearctic Tachinidae (Diptera). *Entomologist's Record and Journal of Variation.* **103:** 23–38.

GODWIN, P.A. & ODELL, T.M. 1984. Laboratory study of competition between *Blepharipa pratensis* and *Parasetigena silvestris* (Diptera: Tachinidae) in *Lymantria dispar* (Lepidoptera: Lymantriidae). *Environmental Entomology* **13:** 1059–1063.

GÖSSWALD, K. 1949. Pflege des Ameisenparasiten *Tamiclea globula* Meig. (Dipt.) durch den Wirt mit Bemerkungen über den Stoffwechsel in den parasitierten Ameisen. *Verhandlungen der Deutschen Zoologischen Gesellschaft* **1949:** 256–264. [Not seen.]

GRADWELL, G.R. 1957. A food plant and a parasite of *Athalia cornubiae* Benson (Hym., Tenthredinidae). *Entomologist's Monthly Magazine* **93:** 12.

GRANT, J.F. & SHEPARD, M. 1983. Biological characteristics of a South American population of *Voria ruralis* (Diptera: Tachinidae), a larval parasitoid of the Soybean Looper (Lepidoptera: Noctuidae). *Environmental Entomology* **12:** 1673–1677.

GREATHEAD, D.J. 1986. Parasitoids in classical biological control. pp 289–318. *In* Waage, J. and Greathead, D. [Eds.] *Insect parasitoids*. London: Academic Press.

GRENIER, S. 1988. Applied biological control with tachinid flies (Diptera, Tachinidae): a review. *Anzeiger für Schadlingskunde, Pflanzenschutz und Umweltschutz* **61:** 49–56.

HAMM, A.H. 1942. Records of bred Tachinidae (Dipt.) chiefly from the Oxford district. *Entomologist's Monthly Magazine* **78:** 191–192.

HAMMOND, H.E. & SMITH, K.G.V. 1953. On some parasitic Diptera and Hymenoptera bred from lepidopterous hosts. Part 1: with a description of *Frontina laeta* Mg., (Dip: Larvaevoridae). *Entomologist's Gazette* **4:** 273–279.

HAMMOND, H.E. & SMITH, K.G.V. 1955. On some parasitic Diptera and Hymenoptera bred from lepidopterous hosts. Part 2: misc. records of Phoridae, Larvaevoridae (Dipt.), Braconidae, Ichneumonidae, and Eulophidae (Hym.). *Entomologist's Gazette* **6:** 168–174.

HAMMOND, H.E. & SMITH, K.G.V. 1957. On some parasitic Diptera and Hymenoptera bred from lepidopterous hosts. Part 3: records of Tachinidae (Dipt.), Braconidae, Ichneumonidae, Encyrtidae, Pteromalidae, Eulophidae and Scelionidae (Hym.). *Entomologist's Gazette* **8:** 181–189.

HARDS, C.H. 1958. Field meeting: Halling, Kent. *Proceedings and Transactions of the South London Entomological and Natural History Society* **1958:** 73.

HASSELL, M.P. 1968. The behavioural response of a tachinid fly (*Cyzenis albicans* (Fall.)) to its host, the Winter Moth (*Operophtera brumata* (L.)). *Journal of Animal Ecology* **37:** 627–639.

HEATH, J. & EMMET, A.M. 1979. *The Moths and Butterflies of Great Britain and Ireland* **9.** London: Curwen Books.

HERREBOUT, W.M. 1966. The fate of eggs of *Eucarcelia rutilla* Vill. (Diptera: Tachinidae) deposited upon the integument of the host. *Zeitschrift für angewandte Entomologie* **58:** 340–355.

HERREBOUT, W.M. 1969. Some aspects of host selection in *Eucarcelia rutilla* Vill. (Diptera: Tachinidae). *Netherlands Journal of Zoology* **19:** 1–104.

HERREBOUT, W.M. & VEER, J. VAN DER 1969. Habitat selection in *Eucarcelia rutilla* Vill. (Diptera: Tachinidae). Part 3: Preliminary results of olfactometer experiments with females of known age. *Zeitschrift für angewandte Entomologie* **64**: 55–61.

HERTING, B. 1960. Biologie der westpaläarktischen Raupenfliegen (Dipt., Tachinidae). *Monographien zur Angewandten Entomologie* **16**: 1-188.

HERTING, B. 1961. Rhinophorinae. *In* Lindner, E. [Ed.]. *Die Fliegen der paläarktischen Region* **64**(e): 1–36. Stuttgart: E. Schweizerbart.

HERTING, B. 1963. Ein ungewöhnlich adaptierter Eilegeapparat bei den Raupenfliegen der Gattung *Phorocera* R.D. (Dipt., Tachinidae). *Stuttgarter Beiträge zur Naturkunde* (Serie A, Biologie) **117**: 1–6.

HERTING, B. 1966. Beiträge zur Kenntnis der europäischen Raupenfliegen (Dipt. Tachinidae). *Stuttgarter Beiträge zur Naturkunde* (Serie A, Biologie) **146**: 1–12.

HERTING, B. 1977. Beiträge zur Kenntnis der europäischen Raupenfliegen (Dipt. Tachinidae). *Stuttgarter Beiträge zur Naturkunde* (Serie A, Biologie) **295**: 1–16.

HERTING, B. 1981. Typenrevisionen einiger paläarktischer Raupenfliegen (Dipt. Tachinidae) und Beschreibungen neuer Arten. *Stuttgarter Beiträge zur Naturkunde* (Serie A, Biologie) **346**: 1–21.

HERTING, B. 1984. Catalogue of Palearctic Tachinidae (Diptera). *Stuttgarter Beiträge zur Naturkunde* (Serie A, Biologie) **369**: 1–228.

HEY, G.L. 1935. A list of parasites bred from Tortrix and Tineid hosts. *Entomologist's Monthly Magazine* **71**: 186–187.

HSIAO, T.H., HOLDAWAY, F.G. & CHIANG, H.C. 1966. Ecological and physiological adaptations in insect parasitism. *Entomologia Experimentalis et Applicata* **9**: 113–123.

KAHRER, A. VON 1984. Das Schlüpfern der Larven von *Elodia morio* (Fall.) (Tachinidae, Diptera) aus ihren mikrotypen Eiern im Darm der Wirtsraupen und unter Künstlichen Bedigungen. *Zeitschrift für angewandte Entomologie* **97**: 95–101.

KAHRER, A. VON 1987. Untersuchungen zur Biologie und Morphologie von *Elodia morio* (Fall.) (Dipt., Tachinidae). *Zeitschrift für angewandte Entomologie* **104**: 131–144.

KEILIN, D. 1944. Respiratory systems and respiratory adaptations in larvae and pupae of Diptera. *Parasitology* **36**: 1–66.

KLOET, G.S. & HINCKS, W.D. 1975. A check list of British Insects. Part 5: Diptera and Siphonaptera. Second edition. *Handbooks for the Identification of British Insects* **11**(5): 1–139.

MATTHEY, W. 1967. The natural history of three oak-feeding tortricids: *Ptycholoma lecheana* (L.), *Pandemis cerasana* (Hb.) and *Batodes angustiorana* (Haw.). *Entomologist* **100**: 115–120.

McALPINE, J.F. 1989. Phylogeny and classification of the Muscomorpha. pp 1397–1518. *In* McAlpine, J.F. [Ed.] *Manual of Nearctic Diptera* **3**. Agriculture Canada.

McALPINE, J.F., PETERSON, B.V., SHEWELL, G.E., TESKEY, H.J., VOCKEROTH, J.R. & WOOD, D.M. 1981 [Eds.] *Manual of Nearctic Diptera* **1**. Agriculture Canada.

MONTEITH, L.G. 1955. Host preference in *Drino bohemica* Mesn. (Diptera: Tachinidae), with particular reference to olfactory responses. *Canadian Entomologist* **87**: 509–530.

MORLEY, C. 1906. On a few Tachinidae and their hosts. *Entomologist* **39**: 270–274.

MESNIL, L.P. 1944–1975. Larvaevorinae (Tachininae). *In* Lindner, E. [Ed.] *Die Fliegen der paläarktischen Region* **64**(g): 1–1435. Stuttgart: E. Schweizerbart.

MESNIL, L.P. 1980. Dexiinae. *In* Lindner, E. [Ed.] *Die Fliegen der paläarktischen Region* **64**(f): 1–52. Stuttgart: E. Schweizerbart.

MUESEBECK, C.F.W. 1922. *Zygobothria nidicola*, an important parasite of the Brown-Tail Moth. *Bulletin of the United States Department of Agriculture* **1088**: 1–9.

NETTLES, W.C., JNR. & BURKS, M.L. 1975. A substance from *Heliothis virescens* larvae stimulating larviposition by females of the tachinid, *Archytas marmoratus*. *Journal of Insect Physiology* **21**: 965–978.

O'HARA, J.E. 1988a. Survey of first instars of the Siphonini (Diptera: Tachinidae). *Entomologica Scandinavica* **18**: 367–382.

O'HARA, J.E. 1988b. Correlation between wing size and position of a hind crossvein in the Siphonini (Diptera: Tachinidae). *Journal of Natural History* **22**: 1141–1146.

O'HARA, J.E. 1989. Systematics of the genus group taxa of the Siphonini (Diptera: Tachinidae). *Quaestiones Entomologicae* **25**: 1–229.

O'HARA, J.E. & COOPER, B.E. 1992. Revision of the Nearctic species of *Cyzenis* Robineau-Desvoidy (Diptera: Tachinidae). *Canadian Entomologist* **124**: 785–813.

PAPE, T. 1987. The Sarcophagidae (Diptera) of Fennoscandia and Denmark. *Fauna Entomologica Scandinavica* **19**: 1–203.

PAPE, T. 1992. Phylogeny of the Tachinidae family-group (Diptera: Calyptratae). *Tijdschrift voor Entomologie* **135**: 43–86.

PARMENTER, L. 1953. Some records of bred Tachinidae. *Entomologist's Record and Journal of Variation* **65**: 29–31.

PELHAM-CLINTON, E.C. 1959. The host of *Billaea carinifrons* (Fallén) (Dipt., Tachinidae). *Entomologist's Monthly Magazine* **95**: 87.

PHILLIPS, M.L. 1983. Parasitism of the common earwig *Forficula auricularia* (Dermaptera: Forficulidae) by tachinid flies in an apple orchard. *Entomophaga* **28**: 89–96.

PHILLIPS, W.M. 1977. Observations on the biology and ecology of the chrysomelid genus *Haltica* Geoff. in Britain. *Ecological Entomology* **2**: 205–216.

PLANTEVIN, G., GRENIER, S., RICHARD, G. & NARDON, C. 1986. Larval development, developmental arrest, and hormone level in the couple *Galleria mellonella* (Lepidoptera-Pyralidae) – *Pseudoperichaeta nigrolineata* (Diptera-Tachinidae). *Archives of Insect Biochemistry and Physiology* **3**: 457–469.

PSCHORN-WALCHER, H. 1969. The host specificity of Tachinidae (Diptera) attacking sawflies (Hym.: Symphyta). *Technical Bulletin. Commonwealth Institute of Biological Control* **12**: 29–36.

RAMADHANE, A., GRENIER, S. & PLANTEVIN, G. 1987. Physiological interactions and development synchronisations between non-diapausing *Ostrinia nubilalis* larvae and the tachinid parasitoid *Pseudoperichaeta nigrolineata*. *Entomologica Experimentalis et Applicata* **45**: 157–165.

RENNIE, J. & SUTHERLAND, C.H. 1920. On the life history of *Bucentes (Siphona) geniculata* (Diptera: Tachinidae), parasite of *Tipula paludosa* (Diptera) and other species. *Parasitology* **12**: 199–211.

RICHARDS, O.W. 1935. Some breeding records of Diptera. *Journal of the Society for British Entomology* **1**: 80–81.

RICHARDS, O.W. 1940. The biology of the Small White Butterfly (*Pieris rapae*) with special reference to the factors controlling its abundance. *Journal of Animal Ecology* **9**: 243–288.

RICHARDS, O.W & WALOFF, N. 1948. The hosts of four British Tachinidae (Diptera). *Entomologist's Monthly Magazine* **84**: 127.

RICHARDS, O.W & WALOFF, N. 1959. *Macquartia brevicornis* Macq. (= *occlusa* Rond.), a new British tachinid fly. *Entomologist's monthly Magazine* **95**: 144.

ROLAND, J., EVANS, W.G. & MYERS, J.H. 1989. Manipulation of oviposition patterns of the parasitoid *Cyzenis albicans* (Tachinidae) in the field using plant extracts. *Journal of Insect Behaviour* **2**: 487–503.

ROGNES, K. 1986. The systematic position of the genus *Helicobosca* Bezzi with a discussion of the monophyly of the calyptrate families Calliphoridae, Rhinophoridae, Sarcophagidae and Tachinidae (Diptera). *Entomologica Scandinavica* **17**: 75–92.

ROGNES, K. 1991. Blowflies (Diptera, Calliphoridae) of Fennoscandia and Denmark. *Fauna Entomologica Scandinavica* **24**: 1–272.

SABROSKEY, C.W. & REARDON, R.C. 1976. Tachinid parasites of the Gypsy Moth *Lymantria dispar*, with keys to adults and puparia. *Miscellaneous publications of the Entomological Society of America* **10**(2): 1–80.

SALT, G. 1968. The resistance of insect parasitoids to the defence reactions of their hosts. *Biological Reviews* **43**: 200–232.

SCHRÖDER, D. 1969. *Lypha dubia* (Fall.) (Dipt.: Tachinidae) as a parasite of the European Pine Shoot Moth, *Rhyaciona buoliana* (Schiff.) (Lep.: Eucosmidae) in Europe. *Technical Bulletin. Commonwealth Institute of Biological Control* **12**: 43–60.

SHAW, M.R. 1979. *Rogas pulchripes* (Wesmael) (Hymenoptera: Braconidae) and other parasites of arboreal *Acronicta* species (Lepidoptera: Noctuidae) at Chat Moss, Manchester. *Entomologist's Gazette* **30**: 291–294.

SMITH, K.G.V. 1961. *Eucarcelia intermedia* Herting (Dipt., Tachinidae) in Britain. *Entomologists's Monthly Magazine* **96**: 117.

SOUTH, R. 1961. *The Moths of the British Isles*. 2 vols. London: F. Warne.

SPERRING, A.H. 1932. Dipterous and hymenopterous parasites bred from Lepidoptera. *Journal of the Entomological Society of the South of England* **1**: 84.

SPOONER, G.M. 1974. *Labigastera forcipata* (Mg.) (Dipt., Tachinidae) new to Britain. *Entomologist's Monthly Magazine* **110**: 85.

161

SPRATT, D.M. & WOLF, G. 1972. A tachinid parasite of *Dasybasis oculata* (Ricardo) and *Dasybasis hebes* (Walker) (Diptera, Tabanidae). *Journal of the Australian Entomological Society* **11:** 260.

STRICKLAND, E.H. 1930. Phagocytosis of internal insect parasites. *Nature* **126:** 95.

THOMPSON, A.C., ROTH, J.P. & KING, E.G. 1983. Larviposition kairomone of the tachinid *Lixophaga diatraeae*. *Environmental Entomology* **12:** 1312–1314.

THOMPSON, W.R. 1915. Sur le cycle évolutif de *Fortisia foeda*, Diptère parasite d'un *Lithobius*. *Compte-Rendu des Séances de la Société de biologie. Paris* **78:** 413–416.

TOWNSEND, C.H.T. 1908. A record of results from rearings and dissections of Tachinidae. *Technical Series. Bureau of Entomology, United States Department of Agriculture* **12:** 95–118.

TSCHORSNIG, H.P. 1985. Taxonomie forstlich wichtiger Parasiten: Untersuchungen zur Struktur des männlichen Postabdomens der Raupenfliegen (Diptera, Tachinidae). *Stuttgarter Beiträge zur Naturkunde* (Serie A, Biologie) **383:** 1–137.

TSCHORSNIG, H.P. 1989. Diagramme zur Flugzeit mitteleuropaischer Raupenfliegen I. Exoristini, Blondeliini. *Mitteilungen. Entomologisher Verein, Stuttgart* **24:** 35–49.

UFFEN, R.W.J. 1961. Miscellaneous notes on Diptera. Part 1: Tachinidae and parasitic Calliphoridae. *Entomologist's Gazette* **12:** 46–49.

VAN EMDEN. see EMDEN

VINCENT, L.S. 1985. The first record of a tachinid fly as an internal parasitoid of a spider (Diptera: Tachinidae; Araneae: Antrodiaetidae). *Pan-Pacific Entomologist* **61:** 224–225.

VOSSBRINCK, C.R. & FRIEDMAN, S. 1989. A 28s ribosomal RNA phylogeny of certain cyclorrhaphous Diptera based upon a hypervariable region. *Systematic Entomology* **14:** 417–431.

WAINWRIGHT, C.J. 1905. Notes on Tachinidae. Number 1. *Entomologist's Monthly Magazine* **41:** 199–207.

WAINWRIGHT, C.J. 1928. The British Tachinidae. *Transactions of the Entomological Society of London* **76:** 139–254.

WAINWRIGHT, C.J. 1932. The British Tachinidae (Diptera). First Supplement. *Transactions of the Entomological Society of London* **80:** 405–424.

WAINWRIGHT, C.J. 1940. The British Tachinidae (Diptera). Second Supplement. *Transactions of the Entomological Society of London* **90:** 411–448.

WALKER, M.G. 1943. Notes on the biology of *Dexia rustica* F., a dipterous parasite of *Melolontha melolontha* L. *Proceedings of the Zoological Society of London.* Series A **113:** 126–176.

WALKER, M.F. 1962. *Degeeria luctuosa* (Meig.) (Dipt., Tachinidae) as a Coccinellid parasite. *Entomologist's Monthly Magazine* **98:** 20.

WALOFF, N. 1987. Observations on the heather beetle *Lochmaea suturalis* (Thomson) (Coleoptera, Chrysomelidae) and its parasitoids. *Journal of Natural History* **21:** 545–556.

WARDLE, R.A. 1914. Preliminary observations upon the life-histories of *Zenilla pexops* B. & B., and *Hypamblys albipicta* Grav. (two previously unrecorded parasites of the Large Larch Sawfly). *Journal of Economic Biology* **9:** 85–104.

WESELOH, R.M. 1980. Host recognition behaviour of the tachinid parasitoid, *Compsilura concinnata*. *Annals of the Entomological Society of America* **73:** 593–601.

WESELOH, R.M. 1983. Effects of multiple parasitism on the Gypsy Moth parasites *Apanteles melanoscelous* (Hymenoptera: Braconidae) and *Compsilura concinnata* (Diptera: Tachinidae). *Environmental Entomology* **12:** 599–602.

WILLIAMS, S.C., ARNAUD, P.H., JNR. & LOWE, G. 1990. Parasitism of *Anuroctonus phaiodactylus* (Wood) and *Vaejovis spinigerus* (Wood) (Scorpiones: Vaejovidae) by *Spilochaetosoma californicum* Smith (Diptera: Tachinidae), and a review of parasitism in scorpions. *Myia* **5:** 11–27.

WINTER, T.G. 1974. New host records of Lepidoptera associated with afforestation in Britain. *Entomologist's Gazette* **25:** 247–258.

WOOD, D.M. 1972. A revision of the New World Exoristini (Diptera: Tachinidae). Part 1: *Phorocera* subgenus *Pseudotachinomyia*. *Canadian Entomologist* **104:** 471–503.

WOOD, D.M. 1987. Tachinidae. pp 1193–1269. *In* McAlpine, J.F. [Ed.]. *Manual of Nearctic Diptera* 2. Agriculture Canada.

WOOD, D.M. & WHEELER, A.G. 1972. First record in North America of the centipede parasite *Loewia foeda* (Diptera: Tachinidae). *Canadian Entomologist* **104:** 1363–1367.

WOODROFFE, G.E. 1953. *Actia antennalis* (Rond.) (Dipt., Larvaevoridae) bred from larvae of *Monopsis rusticella* (Clerk) (Lep., Tineidae). *Entomologist's Monthly Magazine* **89:** 11.

WYATT, N.P. 1986. *Thecocarcelia acutangulata* (Macquart) (Diptera: Tachinidae), new to Britain. *Entomologist's Monthly Magazine* **122:** 203–204.

WYATT, N.P. & STERLING, P.H. 1988. Parasites of the Brown-Tail Moth, *Euproctis chrysorrhoea* (L.) (Lep., Lymantriidae), including two Diptera (Tachinidae, Sarcophagidae) new to Britain. *Entomologist's Monthly Magazine* **124:** 207–213.

ZUSKA, J. 1962. The first instar larvae of the genus *Trixa* Meigen and remarks on the systematics and nomenclature of this genus (Diptera, Larvaevoridae). *Acta Societatis entomologicae Cechosloveniae* **59:** 80–86.

Index

Names in brackets are incorrect (synonyms or old combinations) and are followed by the correct name. Numbers in bold refer to the Species biology section.

Cistogaster 48 **116**
Clemelis 18 **82**
Cleonice 20 **94**
(=cloacellae) Elodia ambulatoria
(=coerulescens) Trixa caerulescens
collaris, Medina 52 **66**
collini Mesnil, Siphona 55 **102**
(=collini Wainwright) Blepharomyia piliceps
(=comata) Carcelia lucorum
Compsilura 17 **69**
comta (=compta), Linnaemya 49 **89**
concinnata, Compsilura 17 **69**
confinis, Aplomya 24 **72**
(=conjugata) Eurithia intermedia
connivens, Eurithia 46 **93**
consobrina, Eurithia 46 **93**
conspersa Meigen, Smidtia 22 **70**
conspersa Harris, Trixa 58 **106**
(=convexifrons) Elodia ambulatoria
costata, Wagneria 59 **111**
(=cotei) Phebellia stulta
(=Craspedothrix) Phytomyptera
cràssicornis, Actia 39 **100**
cristata Fabricius, Siphona 56 **103**
cristata Meigen, Estheria 47 **108**
(=Crocuta) Siphona
cruentata, Winthemia 59 **71**
(=cunctans) Ramonda prunaria
(=curvicauda) Phania funesta
curvinervis, Athrycia 41 **111**
Cylindromyia 27 **119**
CYLINDROMYIINI **118**
Cyrtophleba 16 **112**
Cyzenis 21 **83**

debilitata, Vibrissina 33 **70**
(=decorata) Admontia seria
(=Degeeria) Medina
delecta, Eloceria 35 **95**
(=delicatula) Siphona pauciseta
Demoticus 34 **104**
Dexia 15 **108**
DEXIINAE **106**
DEXIINI **106**
Dexiosoma 16 **106**
(=Digonochaeta) Triarthria
(=diligens) Tlephusa cincinna
diluta, Xylotachina 38 **80**
(=dimano) Eriothrix rufomaculata
Dinera 15 **107**
Dionaea 34 **117**
Diplostichus 18 **63**
(=Discochaeta) Eurysthaea
(=discolor) Meigenia dorsalis
dispar, Macquartia 51 **96**
divisa, Gonia 48 **86**
dorsalis, Meigenia 52 **65**
Drino 24 **77**
dubia, Lypha 50 **90**

Dufouria 19 **113**
DUFOURIINI **113**

(=Echinomya or Echinomyia) Tachina
(=Elfia) Phytomyptera
Eloceria 35 **95**
Elodia **85**
(=Elpe) Campylocheta
Entomophaga 38 **99**
Epicampocera 22 **74**
Eriothrix 20 **109**
Ernestia 20 **91**
ERNESTIINI **91**
(=erucarum) Exorista mimula
(=erucarum of Emden, 1954) Exorista
    tubulosa
Erycia 34 **80**
ERYCIINI **72**
Erycilla 37 **84**
Erynnia 28 **84**
(=Erythrocera) Eurysthaea
Estheria 16 **107**
(=Ethilla) Cadurciella
Eumea 23 **81**
Eurithia 45 **92**
(=Euryclea) Carcelia
Eurysthaea 36 **84**
EUTHERINI **115**
(=Eversmannia or Eversmania) Lypha
(=Evibrissa) Hemyda
(=evonymellae) Eurysthaea scutellaris
(=excavata) Carcelia gnava
excisa, Senometopia 42 **79**
exigua, Microsoma 27 **114**
exoleta, Entomophaga 40 **99**
Exorista 29 **61**
EXORISTINAE **61**
EXORISTINI **61**

fasciata Fallén, Exorista 47 **61**
fasciata Macquart, Rondania 25 **114**
fasciata Meigen, Lophosia 28 **118**
(=fasciata Meigen) Gonia picea
(=fatua) Erycia furibunda
Fausta 45 **92**
fenestrata, Solieria 57 **105**
fera, Tachina 29 **87**
ferox, Nowickia 29 **88**
ferruginea, Erycilla 37 **84**
(=festinans) Erycia furibunda
fimbriata, Platymya 23 **81**
fissicornis, Peribaea 39 **101**
flavipes Robineau-Desvoidy, Hebia 28 **85**
(=flavipes Meigen) Macquartia dispar
(=flavipes Meigen of Emden, 1954)
    Macquartia viridana
floralis, Nemorilla 23 **71**
(=floralis) Meigenia mutabilis
foeda, Loewia 50 **95**

166